STAR STRUCK

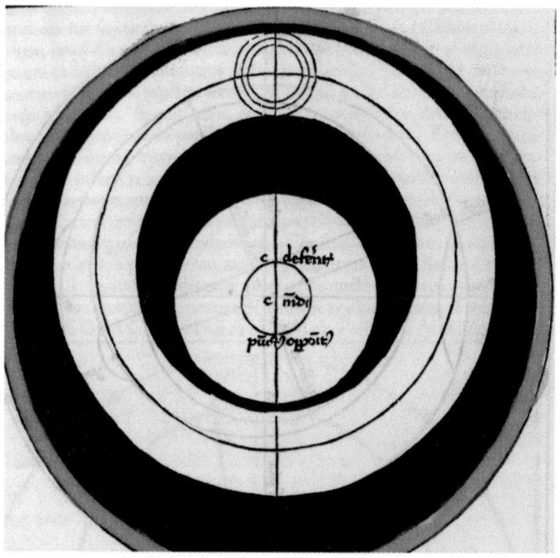

One Thousand Years

of

the Art

and

Science

of

Astronomy

ARMILLÆ ÆQVATORIÆ.

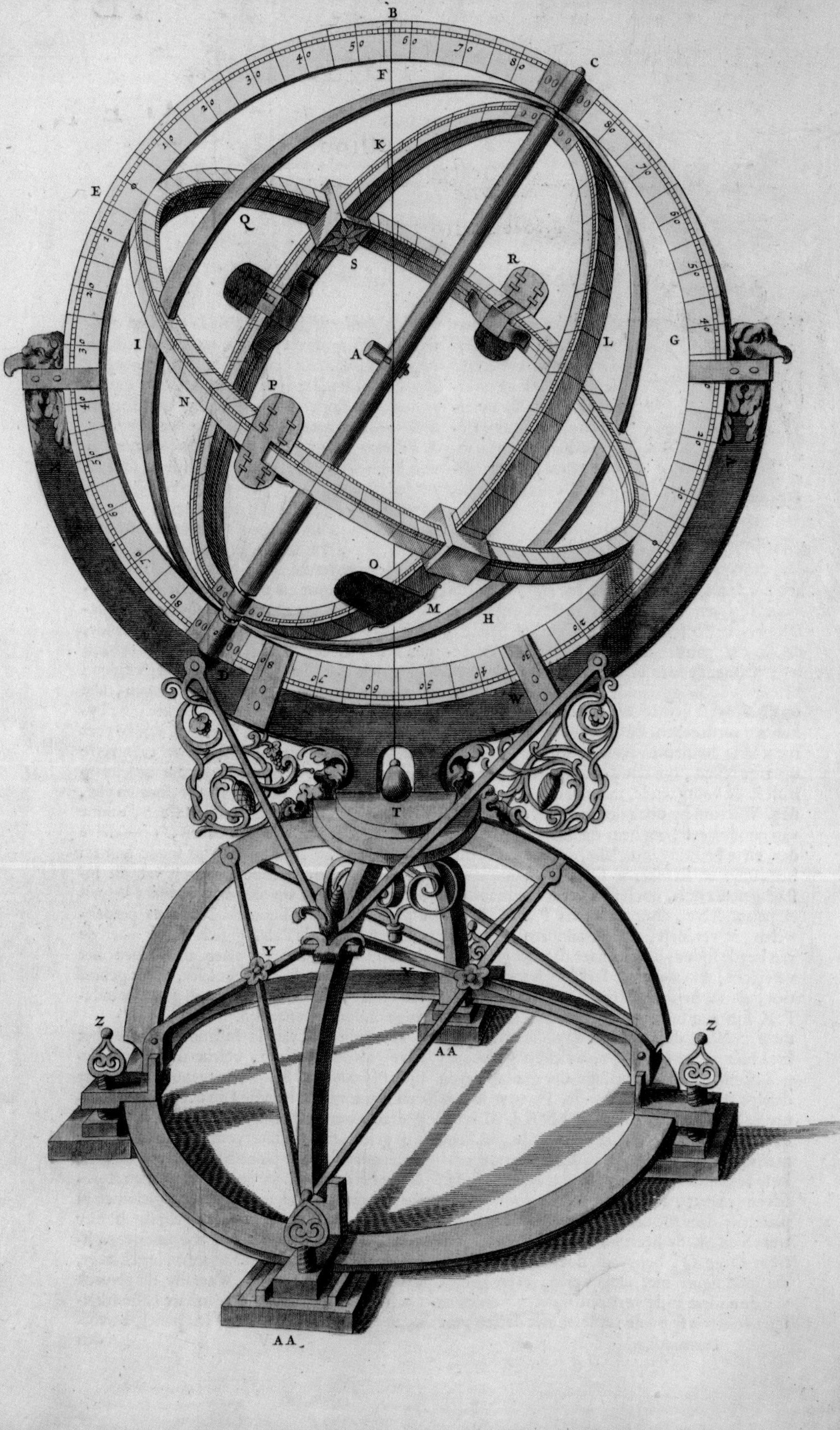

STAR STRUCK

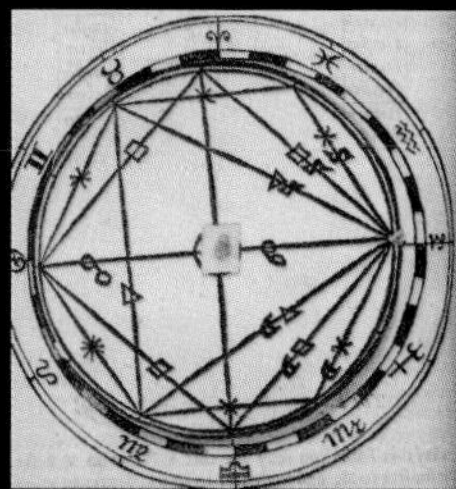

One Thousand Years *of* the Art *and* Science *of* Astronomy

RONALD BRASHEAR
DANIEL LEWIS
with a historical perspective by
OWEN GINGERICH

HUNTINGTON LIBRARY
San Marino
in association with the
UNIVERSITY OF WASHINGTON PRESS
Seattle and London

The Huntington exhibition that inspired this catalog
has been made possible through the generous support of

THE H. RUSSELL SMITH FOUNDATION

THE W. M. KECK FOUNDATION

WELLS FARGO FOUNDATION

RESEARCH CORPORATION

THE BOEING COMPANY

IDA HULL LLOYD CROTTY FOUNDATION IN HONOR AND MEMORY OF
EDWIN AND GRACE HUBBLE

IBM CORPORATION

JUDITH AND BRYANT DANNER

WARREN AND KATHERINE SCHLINGER

ELISE MUDD MARVIN

THE GENCORP FOUNDATION - AEROJET

SMITHSONIAN INSTITUTION LIBRARIES

. . .

Printed in China

Huntington Library edition

ISBN 0-87328-191-8 (ppb) ISBN 0-87328-186-1 (cloth)

University of Washington Press edition

ISBN 0-295-98097-4 (ppb) ISBN 0-295-98096-6 (cloth)

TITLE PAGE AND BACK COVER:
An observational zodiacal armillary sphere, from Jean Blaeu's work of 1663,
Le grand atlas, ou cosmographie Blaviane...

FRONT COVER:
In 1610 Galileo published the first images of the Moon as seen through a telescope. His depictions of the waning gibbous phase and the last quarter of the Moon from *Sidereous Nuncius* (Starry messenger) are juxtaposed with a detailed modern image shot at Mount Wilson in Southern California.

Contents

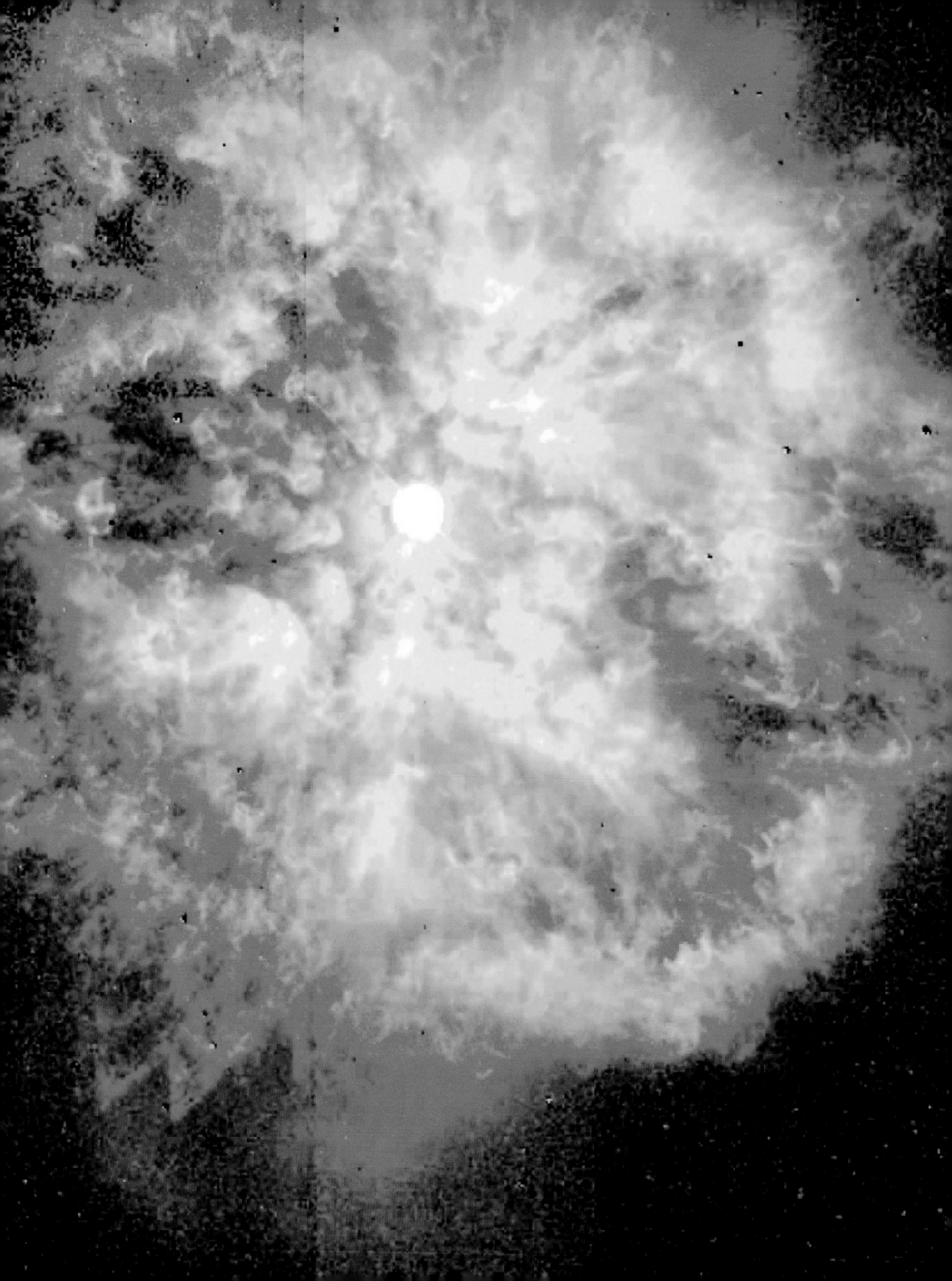

STAR STRUCK | *A Historical Perspective*

THE SKY ABOVE—eternal and serene, yet intricate and mysterious—has piqued both reverence and human curiosity for at least as long as records survive. Bone and clay, stone, papyrus, and vellum, have all documented the human quest to understand the motions and structures of the heavens. In the past millennium parchment has given way to paper, the scribal hand to printing, the naked eye to glass giants and photography, and now, on the threshold of a new age, our observations reside ever more frequently in strangely intangible electronic data banks.

For the woman or man on the street, current concepts of the universe are molded by postcards from space, those magnificent close-ups of Saturn's rings, the haunting otherworldly fingers of the Eagle Nebula, or the wispy polka dots of the Hubble deep fields. Our grander notions of the vast scale of creation are more difficult to picture. This stupefying scale—in both space and time—is an invention of the past century, with fine-tuning scarcely a decade old. These achievements of human intellect are tantalizingly abstract for simple visions. The heliocentric system, the Milky Way spiral, the Big Bang—these concepts cannot be brought into an exhibit as original artifacts, especially when they are creations of contemplation.

Here is where books and manuscripts, and the great libraries that preserve them, play their essential role. As physical evidence from years or centuries past, these paper artifacts can celebrate intellectual achievements as abstract as Copernicus's revolutionary step from a geocentric to a heliocentric blueprint of the cosmos—a "theory pleasing to the mind," as he expressed it.

This exhibition brings together treasures largely drawn from the Huntington Library's own collections. Some of the items are one-of-a-kind, such as Edwin Hubble's logbook from the Mount Wilson Observatory's 100-inch Hooker reflector, marked to show his discovery of the first Cepheid variable found in the Andromeda nebula. The star's rhythmic pulsations betrayed its extraordinary absolute brightness, which ultimately established its incredibly far-flung locale and the first rung of the extragalactic distance ladder. Other items on exhibit are unique by association, such as the second edition of Copernicus's *De revolutionibus*, which was owned by Hubble.

A famous letter in German from Albert Einstein to George Ellery Hale (the founder of Mount Wilson and Palomar Mountain observatories) requesting an observational test of his new theory of general relativity is unique, as are several additional fascinating letters and manuscripts in the collection. Perhaps fearing that Hale had never heard of Einstein (who in 1913 had not yet achieved universal name recognition), the Zurich institute's director appended a charming English postscript: "Many, many thanks for a friendly reply to the professor Dr. Einstein, my honorable Colleague of the Polytechnical school."

This dramatic NASA Hubble Space Telescope picture of the energetic star WR124 reveals that it is surrounded by hot clumps of gas being ejected into space at speeds of more than 100,000 miles per hour. Vast arcs of glowing gas around the star make it resemble an aerial fireworks explosion. This extremely rare, short-lived and massive, central superhot star, known as a Wolf-Rayet star, is going through a violent, transitional phase characterized by the fierce ejection of mass. The blobs may result from the furious stellar wind that does not flow smoothly into space but has instabilities which make it clumpy. The star is 15,000 light-years away, located in the constellation Sagittarius. The picture was taken with Hubble's Wide Field Planetary Camera 2 in March 1997. The image is false-colored to reveal details in the nebula's structure. Credit: Yves Grosdidier (University of Montreal and Observatoire de Strasbourg), Anthony Moffat (University of Montreal), Gilles Joncas (Université Laval), Agnes Acker (Observatoire de Strasbourg), and NASA.

Other items, once common, are today precious heirlooms. Thomas Digges's *Prognostication everlasting: A perfit description,* which in 1576 carried an English translation of a few key chapters of Copernicus's treatise and which for the first time displayed a diagram with the stars scattered out toward infinity, probably flooded the bookshops in an edition of a thousand copies. Today only about a dozen examples survive; the only other copy in America is in the Folger Shakespeare Library in Washington. But the advantage of printing was that the press stamped out multiple copies, and remarkable ideas could be widely disseminated. Galileo's *Dialogo* or *Dialogue on the Two Chief World Systems* (1632), the vernacular book that got him into trouble with the Inquisition, was also published in a thousand copies, but its readers quickly recognized the brilliant defense of Copernican astronomy as an epoch-making book, so hundreds survive. The *Dialogue* remains a valuable volume, but within the range of many rare book collectors.

For well over half of the last millennium, Ptolemy's earth-centered planetary system held sway, and a substantial number of items on exhibit reflect that older view of the cosmos. Included is a scarce copy of Georg von Peurbach's *Theoricae novae planetarum* (1474)—not actually a new theory of the planets, but a new *Theorica,* that is, a new theoretical textbook. The work was published by the astronomer Johann Mueller who was Peurbach's junior colleague and better known as Regiomontanus, the world's first science printer. Impressed by the power of the new medium, Regiomontanus set out to print a long list of astronomy and mathematics titles, and before his untimely death in 1476 he actually issued half a dozen of them, including the now-rare *Astronomicon* by Marcus Manilius, which is also on display.

The most spectacular and most valuable printed book in the exhibition, Petrus Apianus's *Astronomicum Caesareum* (1540), establishes the high water mark of geocentric astronomy. Its complex layering of movable disks (the so-called volvelles) model the Ptolemaic universe and allow the skilled user to predict the position of the planets by manipulating the paper instruments as analog computers. Johannes Kepler later decried the immense effort that had gone into depicting a cosmological system on its way out, but in 1540 Apianus had little way of knowing what the next decades would bring.

As the millennium advanced past its sixth century, a crucial technological advance, the telescope, was applied to the heavens and brought changes so overwhelming that the exhibit quite rightly moves to a new subheading here: "Technologies of Observing" supplants the ongoing tasks of "Representing the Heavens." Galileo's pioneering *Sidereus Nuncius* (Starry messenger) of 1610, the second most valuable printed book on view, proclaimed what the new "optick tube" could do in the hands of a skilled technician and observer who essentially converted a toy into a scientific instrument. Homage to the ever more perspicacious telescopes developed from the seventeenth century to modern times includes Isaac Newton's novel reflector (which won him election to the Royal Society) and William Herschel's forty-foot behemoth. These led, with major leaps of technology, to the extraordinary instruments that Hale had built for Southern California's own Mount Wilson Observatory, all represented in the exhibition.

Armed with these giant eyes, astronomers have not only made major strides in modeling the universe, but whole new vistas have been opened by their extended vision. In Galileo's hands the Moon yielded its long-held secret. No longer could our satellite be viewed as a perfect orb of crystalline quintessence, for it became an earthlike body of mountains and plains. The heady notion of a plurality of worlds spawned a host of proposals and fantasies regarding life elsewhere, often well informed, always speculative. In his *Celestial Worlds Discover'd* (1698), Christiaan Huygens described why the inhabitants of Saturn would have hemp:

> If their Globe is divided like ours, between Sea and Land, as it's evident it is, we have great reason to allow them the Art of Navigation, and not proudly ingross so great, so useful a thing to our selves. And what a troop of other things follow from this allowance? If they have Ships, they must have Sails and Anchors, Ropes, Pullies, and Rudders....

Galileo made the first advance in deciphering the Milky Way, demonstrating that its milkiness resulted from the confluence of stars unresolved by the naked eye. Over a century would pass before Thomas Wright described the spatial arrangement of stars that produced the luminous band circling the nocturnal heavens. His splendidly illustrated, and at times bizarre, account appeared in 1750 as *An Original Theory or New Hypothesis of the Universe*. The English astronomer William Herschel owned a copy and was challenged to try to attach a distance scale to Wright's proposal. At the same time he added his own speculations about the faint nebulae that he found by the thousands, which he called "island universes." His conjectures and those of the Earl of Rosse, who discovered the spiral structure of the nebulae, finally reached fruition in Edwin Hubble's explorations of the realm of the nebulae. Hubble not only established their great distances but demonstrated that the fainter they appeared, the faster they were rushing away from the Milky Way, thereby founding observational cosmology.

Because of the exponential increase in the number of astronomers and of telescopes—both earthbound and in space—more pages of astronomical research have been written in the past fifty years than in the previous 950 years of the millennium. Yet apart from some special images from space, the exhibition has nary an item from the past five decades. The letter from Einstein and the working notes from Hubble almost carry the entire twentieth century alone. More recent research still lacks the patina of rarity and collectibility brought by age; our perspective on the most modern discoveries remains somewhat unfocused.

As Isaac Newton wrote to his lesser rival, Robert Hooke, "If I have seen further, it is by standing on ye shoulders of giants." The Huntington Library exhibition *Star Struck* highlights the legacy of giants—the Galileos of each age. But also contributing to the march of science there are, besides the thoughtful observers and men of scientific genius, those who through the artistry and imagination of their books informed and inspired their contemporary non-specialist audience. Among them are Petrus Apianus, John Wilkins, Bernard Fontenelle, Benjamin Martin, and Ormsby McKnight Mitchel, illustrious popularizers from the sixteenth to the nineteenth centuries who brought astronomy to kings and commoners alike. That trail led to the science-fiction pioneer H. G. Wells, whose visionary *The First Men in the Moon* (1901) joins examples from the others, all appropriately represented in a work subtitled *The Art and Science of Astronomy*.

The treasured works illustrated here range from the brilliantly but forbiddingly technical—such as Ptolemy's *Almagest* (in the form of a 1279 manuscript as well as the first printed edition of 1515), Kepler's *Astronomiae pars optica* (The Optical part of astronomy)(1604) or Laplace's *Mécanique céleste* (Celestial mechanics) (1802)—to the luminously accessible—the atlases of Bayer (1601) and Dopplemayr (1742) or Galileo's *Dialogo*. Some, like Newton's *Principia* (1687), are unquestioned landmarks in human thought, whereas others are charming antiques that decorate the road to modern science. Altogether they bear witness to the long, continuing cultural impact of the heavens—hoary and still mysterious, yet beckoning and ever challenging.

Owen Gingerich
Harvard-Smithsonian Center for Astrophysics

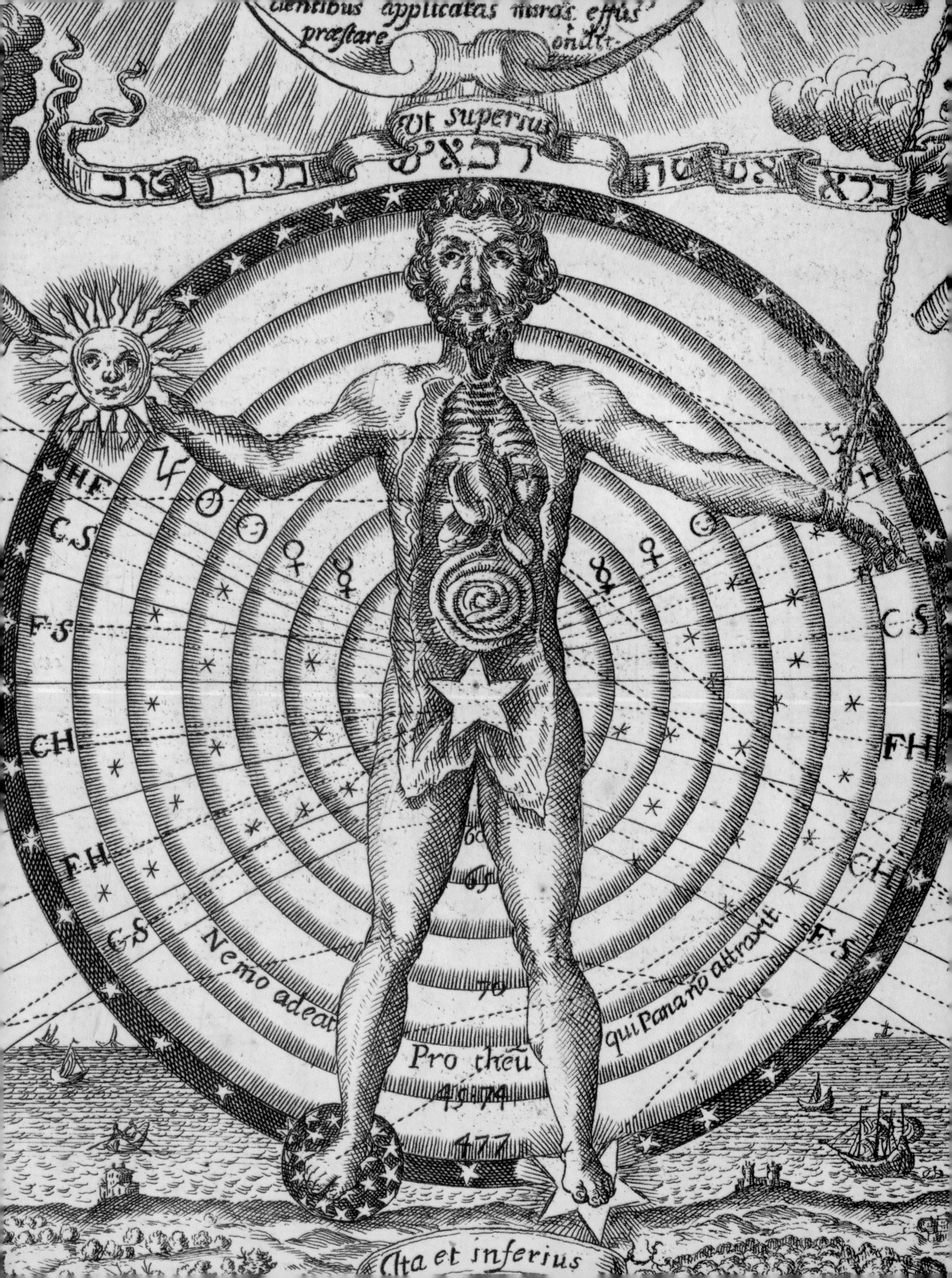
præstare
Vt superius
HF
CS
FS
CH
FH
GS
Nemo adeat
qui Pana nō attraxit
60
65
70
Pro theū
477
Ita et inferius

I

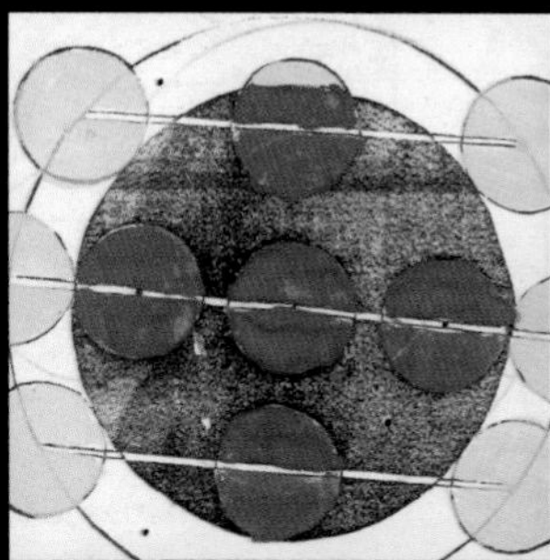

REPRESENTI
THE
HEAVENS

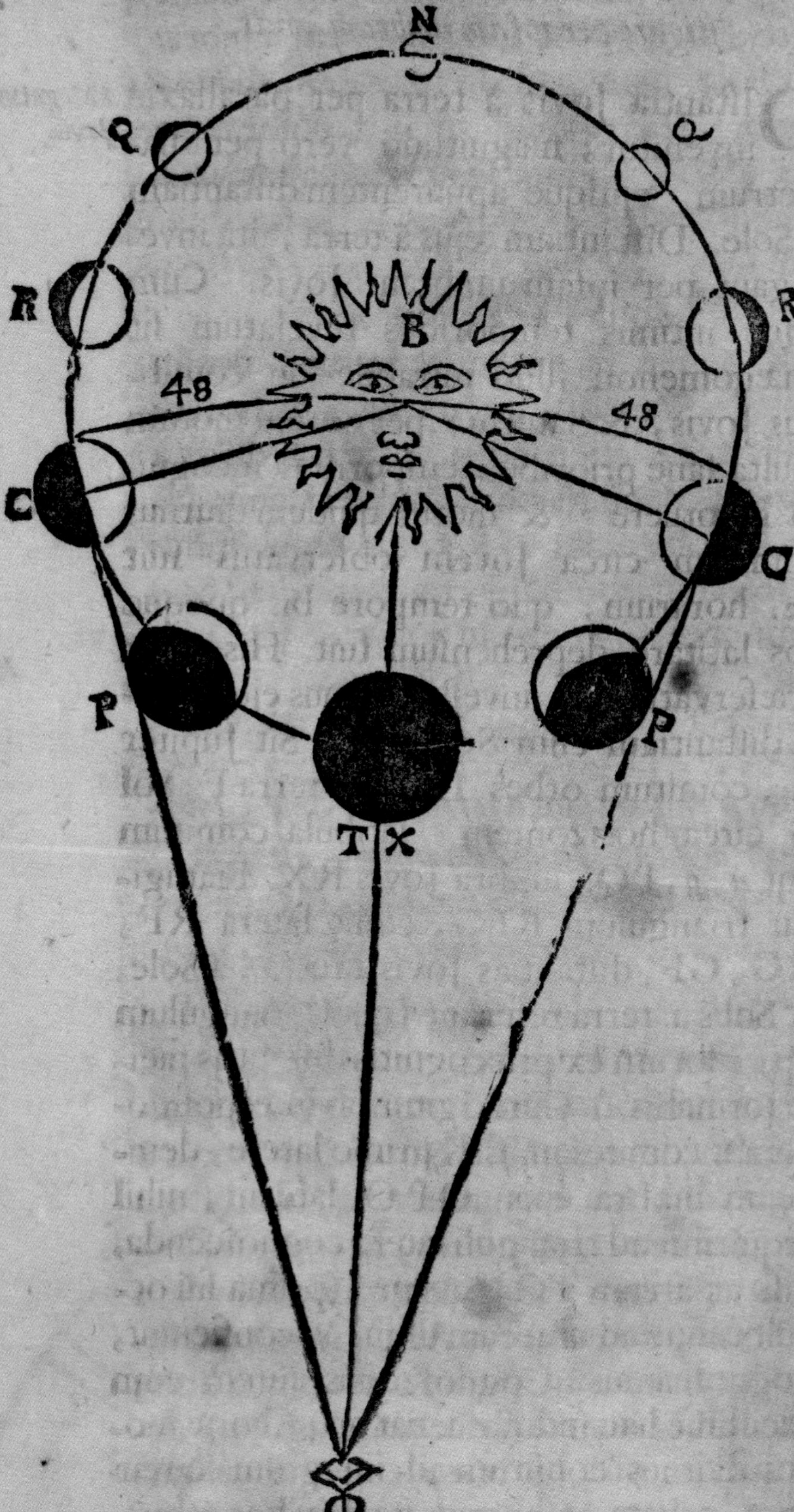

Augis puncto, quasi plena, & cosen-

ONE | ASTROLOGY

THE HEAVENS HAVE FASCINATED HUMANS from earliest times. Astrology attempted to explain and predict the influence of the heavens on nature and on health and behavior. Astrology originated in approximately 1000 B.C.E. in Babylonia, where the practice was used as a form of divination, especially for predicting the course of events for rulers. It also flourished in the east, most notably in the seventh and eighth centuries, where it became more mathematically refined as astrologers and mathematicians tried to plot planetary positions with greater and greater accuracy.[1] As a result the boundaries between astrology and astronomy have long been fuzzy.

Astrology first came to the western world through Islamic teachings and practices in approximately 1200 C.E., and its influence rapidly became widespread in the west. The great English poet Geoffrey Chaucer (1340?–1400) wrote on astrology, including a witty account of the conjunction of Venus and Mars, couched in the form of a somewhat indecent love story. During the Renaissance, astrology became part of a concept known as cosmography, which was defined in the widest possible way as a study of the universe—both heaven and Earth, and the interplay between the two. Astrology was one of four elements of cosmography, the other three being astronomy, geography, and chorography (the description and study of a particular place, whether region, city, ocean, or island).

The first book printed—and thus widely distributed—on astrology and astronomy was Marcus Manilius's *Astronomicon*, published at the end of 1473 or in early 1474 at the Nuremberg press of the great natural philosopher and printer Johann Mueller (also known as Regiomontanus). Manilius lived in Rome at the start of the first century C.E. and is known to us only through this work, intended to instruct students in both astronomy and astrology. The *Astronomicon* is an incomplete poem; only the first five books survive. Books I and V treat questions of astronomy proper: the sphere, zodiacal and other constellations, comets, and the fixed stars; while Books II to IV relate to purely astrological matters. Regiomontanus's printing press was the first devoted exclusively to the diffusion of works of science, and he published other important works on astronomical topics, such as Georg von Peurbach's 1474 work *Theoricae novae planetarum* (New theory of the planets).

Before the Scientific Revolution of the sixteenth and seventeenth centuries, people considered astrology and astronomy to be related and intertwined parts of the study of the mysterious heavens. The Revolution created a greater separation between the two areas of learning, but astrology still continued to be embraced scientifically; the renowned astronomers Johannes Kepler and Galileo Galilei, for

Athanasius Kircher. *Ars magna lucis et umbrae* (3rd edition. Amsterdam: Joannem Janssonium, 1671).

PRECEEDING SPREAD: Detail showing supposed celestial effects on the body.

OPPOSITE: View of the phases of Venus as seen from the Earth. (Huntington Library: RB 487000.1068)

OPPOSITE
Athanasius Kircher. *Ars magna lucis et umbrae* (3rd edition. Amsterdam: Joannem Janssonium, 1671).

TOP: This early astrological chart describes the effects of the celestial position of the planets and stars on different parts of the human body—the eyes, intestines and heart, for instance.

BOTTOM: Astrologers thought that the position of the zodiac and the time of the year had an influence on the workings and health of the body. (Huntington Library: RB 487000.1068)

example, cast horoscopes for their patrons. Astrology, a topic of debate for hundreds of years, has been persistently viewed by some over the centuries as a pseudoscience and by others as solidly grounded truth capable of providing meaningful guidance. This dichotomy is evidence of the difficulties in understanding the nature of the heavens, which is due both to the scientific complexity of the extraterrestrial world as well as to the fact that our understanding of these phenomena, while better and better, remains imperfect. "Astrology… has been dead so long that it no longer stinks; perhaps because it is embalmed in the writings of so many men that were eminent in their day," wrote one observer at the start of the twentieth century. But this same observer—a physician named Charles Mercier—also thought that the planets had a definite influence over health. "Those under the jurisdiction of Luna," he notes, "are tall, pale, good-looking, with light hair and eyes, and with becoming beards. When well affected, they are ingenious, subtle, sincere, open, honest and well mannered; when ill affected, they are stupid even to fatuity, timid and restless."

The use of astrology as a predictive tool was grounded in the increasing accuracy with which astronomers were able to predict such cosmic events as eclipses. This success suggested that future events of a more personal nature could also be predicted. Because the universe was supposedly perfect (spherical in nature, unchanging in its essential qualities, or so observers thought), it was viewed as an omnipotent presence in some ways. Various thinkers considered astrology a useful predictive tool. The English writer Henry Coley (1633–1707) laid out theories of "weather astrology" (the role of the heavens on the weather and the ability to predict it) and "natal astrology" (the preordained influences of the stars on people, based on the sign under which they were born), in a popular work published in 1676, entitled *Clavis Astrologiae Elimata, or, a Key to the Whole Art of Astrology*. Even the great astronomer Claudius Ptolemy (who lived ca. 100–180 C.E.) wrote a textbook of astrology. In his four-part work, entitled *Tetrabiblos* (also known as the *Quadripartitum*) and first published in 1519, Ptolemy, a careful analyzer, describes the influence of heavenly bodies as strictly physical. Observing the influences of the nearby Sun and Moon (on tides, the seasons, and so forth), he extrapolated that the other bodies in the sky could also exert some physical sway over the Earth. By careful observation of these effects, he thought, it would be possible to construct a system that, although not mathematically certain, would enable one to make certain useful predictions.

Astrology has also played a particularly important role in historic perceptions of health and medicine. For several centuries following its introduction into the west, many sages and theologians agreed that the heavens had a causative effect on people. The seventeenth-century German writer and scholar Athanasius Kircher (1601–1680) argued in his 1646 work *Ars magna lucis et umbrae* (The great art of light and shadow) that light was the "attracting magnet of all things" and thus believed that anything which emitted light—the Sun, Moon, and stars—had an influence on the body. But astronomers and philosophers disagreed strongly about the ability of the stars to predict future events.

After approximately 1700, the English scientist and mathematician Isaac Newton's (1642–1727) views of the motions of heavenly bodies had become solidly established and astrologywas relegated to the status of superstition for about two hundred years. The dawn of the twentieth century, however, saw a revival of popular interest in the paranormal. The now-ubiquitous newspaper horoscopes first appeared in the 1930s, and despite regular refutations of astrology by astronomers and other scientists, the concept retains a strong hold on the popular imagination. *DL*

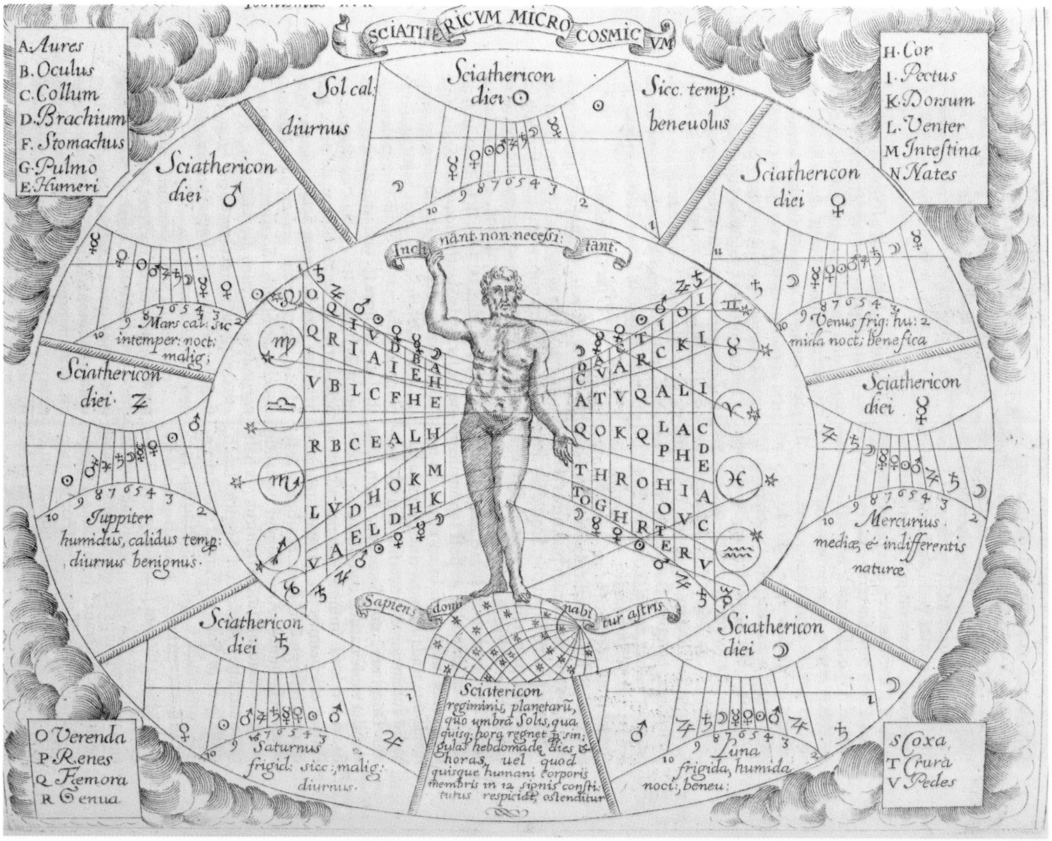
SCIATHE RICVM MICRO COSMIC VM
A. Aures
B. Oculus
C. Collum
D. Brachium
F. Stomachus
G. Pulmo
E. Humeri
H. Cor
I. Pectus
K. Dorsum
L. Venter
M Intestina
N Nates
Sciathericon diei
Sol cal: diurnus
Sicc. temp: beneuolus
Mars cal: sic intemper: nocti: malig:
Venus frig: hu: mida nocti: benefica
Iuppiter humidus, calidus temp: diurnus benignus.
Mercurius mediæ, & indifferentis naturæ
Saturnus frigid: sicc: malig: diurnus.
Luna frigida, humida nocti; beneu:
Inclinant non necessitant.
Sapiens dominabitur astris
Sciatericon regiminis planetarũ, quo umbra Solis, quaquisq; hora regnet & singulas hebdomadæ dies & horas, uel quod quisque humani corporis membris in 12 signis constitutus respicidt, ostenditur
O Verenda
P Renes
Q Femora
R Genua
S Coxa
T Crura
V Pedes

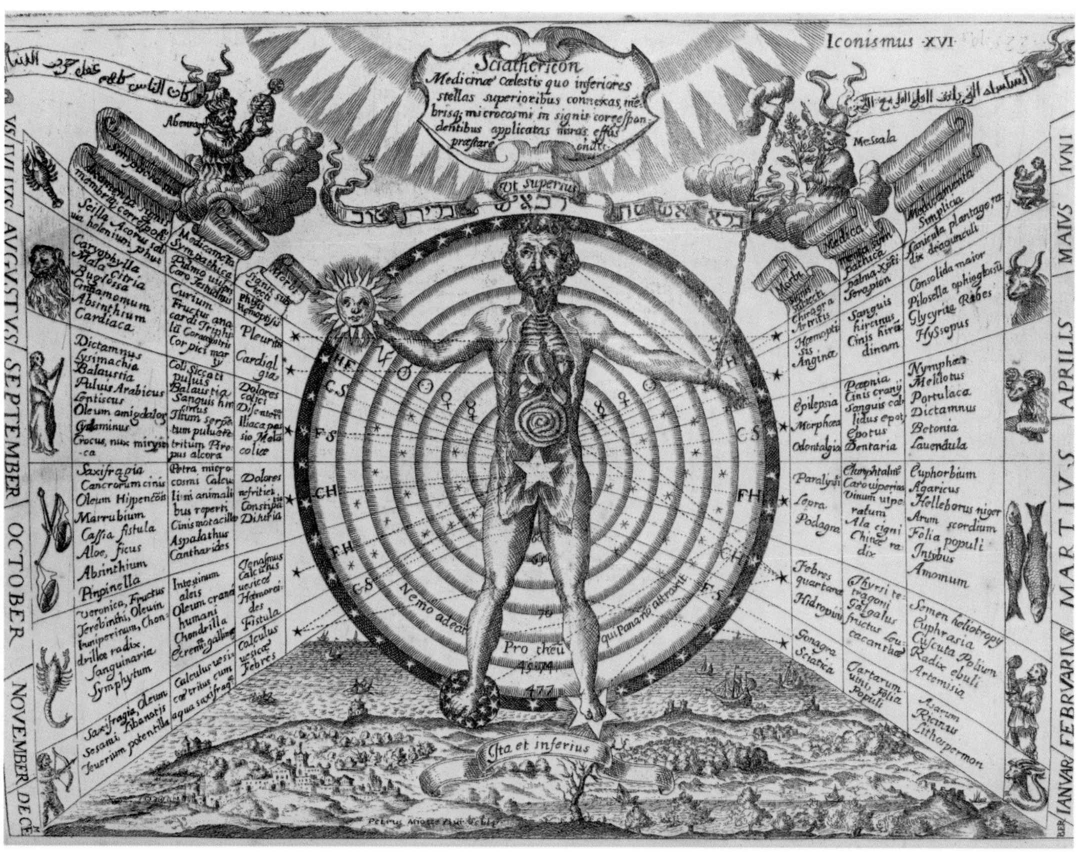
Iconismus XVI
Sciathericon
Medicinæ cœlestis quo inferiores stellas superioribus connexas, mebrisq; microcosmi in signis correspondentibus applicatas miros effectus præstare ostendit
ut superius
Ita et inferius
Abenragel
Messala
IVLIVS AVGVSTVS SEPTEMBER OCTOBER NOVEMBER DECE
IVNI MAIVS APRILIS MARTIVS FEBRVARIVS IANVAR
Nemo adeat qui Pana ter abnegabit

ne alicuius signi nemini iunctus ⁊ ꝑcusserit alteꝝ signum lumine suo quisquis planetarum fuerit dignior lumine illo erit coniunctus ei: lꝫ planeta qui fuerit in primo signo non videat eum.

¶ Expositio recessionis vel separationis planetaꝝ.

Expositio recessionis vel separationis planetaruꝫ est vt preterea planeta leuior alium pōderosiorem: ⁊ incipiat habere plus gradus tam in aspectu q̃ꝫ in ↄiunctione. nam aspectus est a signo in signum. Coniunctio autem dicitur a gradu in gradum: ⁊ hec sciētia est Messabalah: idest quem deus voluit magistrum.

¶ De translatione luminis.

Trāslatio luminis a planeta in planetam est vt separetur planeta leuior ab alio ponderosiori ⁊ iungatur alteri: tunc quasi ↄiungit eos ⁊ defert naturam primi ad alteruꝫ cui iungitur. Cuius rei exemplar est vt esset ascendens virgo: ⁊ fieret interrogatio de ↄiugio: ⁊ esset luna in decimo gradu signi geminoꝝ: ⁊ mercurius in octauo gradu leonis: ⁊ Juppiter in 13° gradu piscis: erat dominus ascendentis qui erat significator interrogationis non aspiciebat Jouem qui est dñs domus coniugij: q2 erat in octauo signo ab eo. Asperi ergo lunam quam. s. inueni in. 10. gradu geminoꝝ separatam a mercurio ⁊ iunctam Joui: deferebat enim inter vtrosqꝫ lumen: ⁊ hic significauit effectum rei: id est acceptionem mulieris ꝑ manus legatoꝝ ⁊ inter vtrosqꝫ discurrentium.

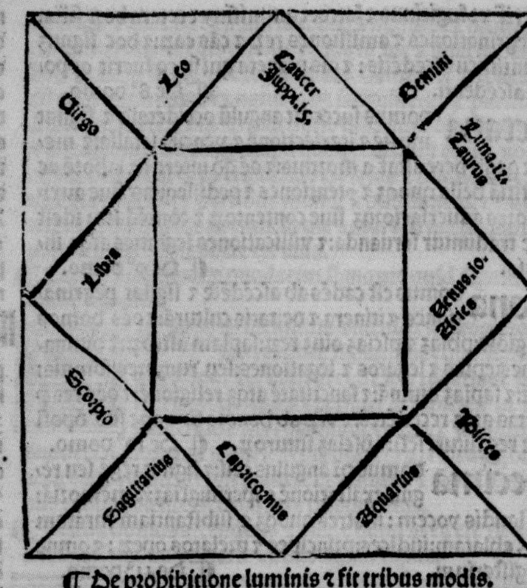

¶ De probibitione luminis ⁊ fit tribus modis.

Almana idest prohibitio fit tribus modis: quorum vnus dicitur abscisio luminis: ⁊ hic fit quādo inter dñm ascendentis ⁊ dominum quesite rei fuerit planeta aliquis in paucioribus gradibus vnius eoꝝ ⁊ fuerit coniunctio cum eodem anteq̃ꝫ fiat ↄiunctio cum dño rei: cuius exemplum est vt esset ascēdens virgo: ⁊ interrogatio fieret de ↄiugio: ⁊ mercurius dñs ascēdentis qui est significator interrogātis in 10° gradu signi cācri: ⁊ Juppiter dñs domus septime qui est significator spōse in. 15. gradu signi piscis: ⁊ mars erat in. 13. gradu arietis: abscidebat ergo mars lumen mercurij a Joue: ⁊ erat mars in. 8. signo s. substantie mulieris: significauit ꝙ destructio huius rei fieret ex descriptione dotis.

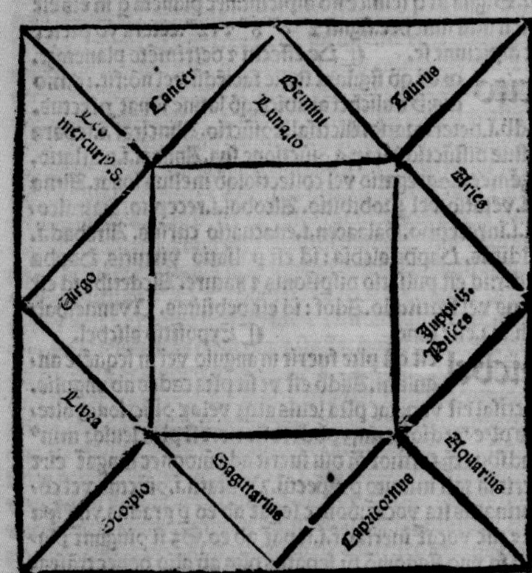

¶ De coniunctione luminis planetaꝝ.

Coniūctio luminis est quando dñs ascendentis ⁊ dñs quesite rei: iungunt planete ponderosiori se: qui coniungat eoꝝ fortitudinem atqꝫ lumen: ⁊ accipiat eoꝝ naturas. verbi gratia. quedam interrogatio fuit de rege. vtrum acquireret regnum an non: ⁊ erat ascēdens signum libre: cuius domina venus que erat significatrix interrogationis in. 10. gradu signi arietis: ⁊ luna domina domus regie: que significat regnum in. 12. gradu signi tauri non aspicientes se: ⁊ erat Juppiter in. 15. gradu signi cancri in angulo celi in domo. s. regia: ⁊ luna atqꝫ Venus iungebantur ei: coniungebat ergo Juppiter lumen: idest radios ambaruꝫ in loco quesite rei: id est in loco regni: ⁊ significat acquisitionem regni per manus cuiusdā ducis siue episcopi: vel ꝑ manus alicuius viri dilecti cui ambo planete libenter tribuant ascensum.

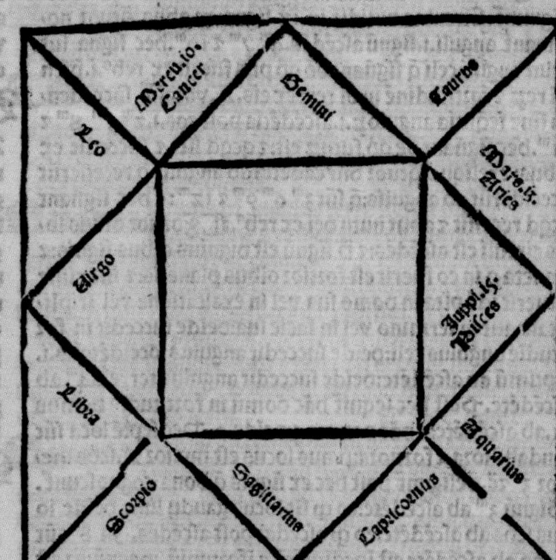

¶ Scđs modus est vt planeta leuis ⁊ alter pōderosior sint ambo in vno signo ⁊ sit tertius inter eos in eodem signo petens coniunctionē ponderosioris: hꝯ aufert ↄiunctioneꝫ primi: cuius exemplum est vt esset ascendens cancer: ⁊ interrogatio fieret de ↄiugio: ⁊ Luna in. 8. gradu geminorū ⁊ mars in gradu. 10. ꝑdicti signi. Saturnus vo in. 12. gradu eius

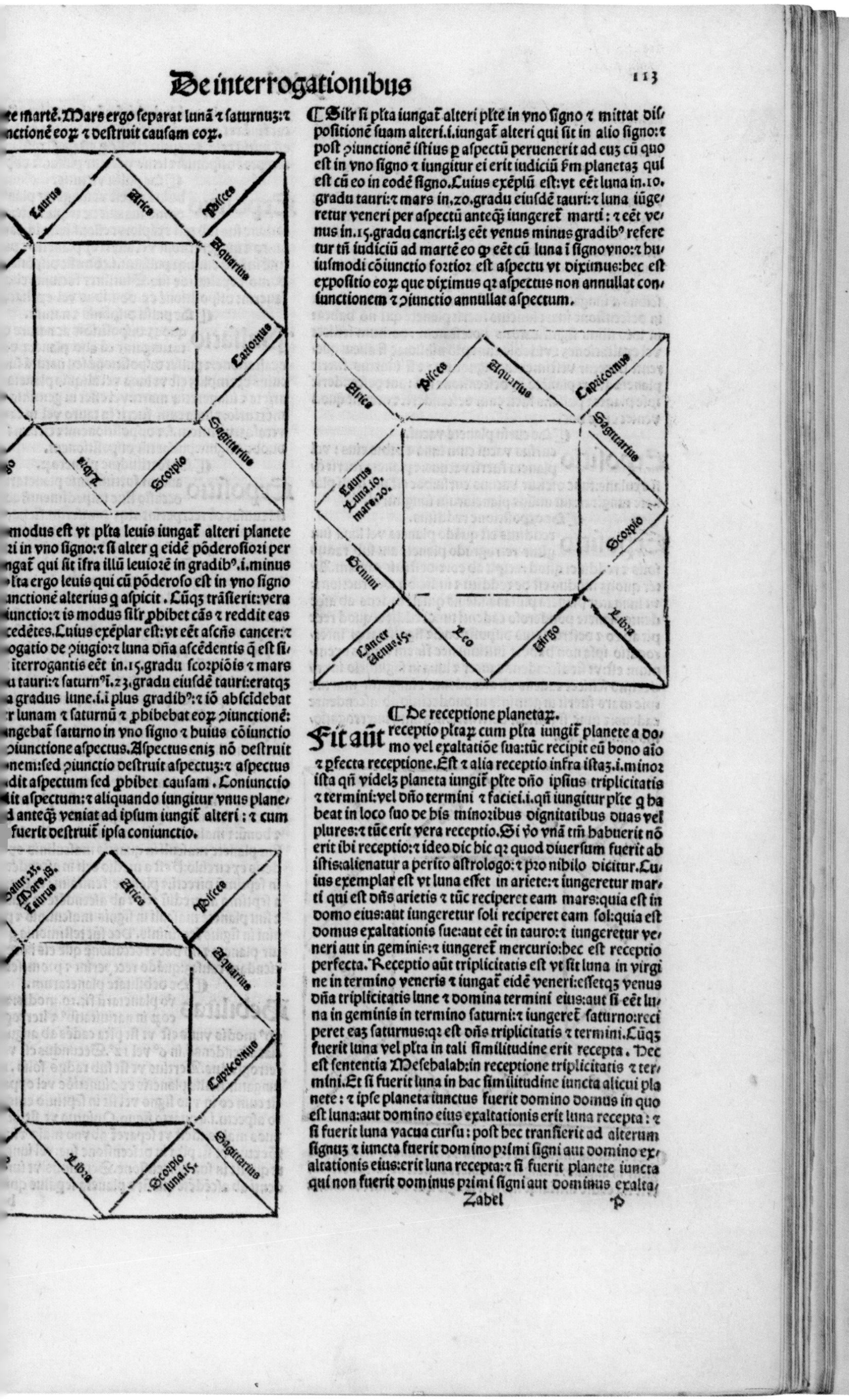

De interrogationibus 113

¶ Sĩlr si plãta iungat alteri plãte in vno signo τ mittat dispositionē suam alteri.i.iungat alteri qui sit in alio signo: τ post ꝯiunctionē istius p aspectū peruenerit ad cuȝ cū quo est in vno signo τ iungitur ei erit iudiciū ꝓm planetaȝ qui est cū eo in eodē signo. Cuius exēplū est: vt eēt luna in.10. gradu tauri: τ mars in.20. gradu eiusdē tauri: τ luna iūgeretur veneri per aspectū anteꝗ̃ iungeret marti: τ eēt venus in.15. gradu cancri: lȝ eēt venus minus gradibꝰ referetur tñ iudiciū ad martē eo ꝙ eēt cū luna i signo vno: τ huiusmodi cōiunctio fortior est aspectu vt diximus: hec est expositio eoꝝ que diximus qꝛ aspectus non annullat coniunctionem τ ꝯiunctio annullat aspectum.

Taurus Luna.10. mars.20.

¶ De receptione planetaꝝ.

Fit aūt receptio pltaꝝ cum plta iungit planete a domo vel exaltatiōe sua: tūc recipit eū bono aīo τ pfecta receptione. Est τ alia receptio infra istaȝ.i.minor ista qñ videlȝ planeta iungit plte dño ipsius triplicitatis τ termini: vel dño termini τ faciei.i.qñ iungitur plte ꝗ habeat in loco suo de his minoribus dignitatibus duas vel plures: τ tūc erit vera receptio. Si vo vnā tñ habuerit nō erit ibi receptio: τ ideo dic hic qꝛ quod diuersum fuerit ab istis alienatur a perito astrologo: τ pro nihilo dicitur. Cuius exemplar est vt luna esset in ariete: τ iungeretur marti qui est dñs arietis τ tūc reciperet eam mars: quia est in domo eius: aut iungeretur soli reciperet eam sol: quia est domus exaltationis sue: aut eēt in tauro: τ iungeretur veneri aut in geminis: τ iungeret mercurio: hec est receptio perfecta. Receptio aūt triplicitatis est vt sit luna in virgine in termino veneris τ iungat eidē veneri: essetqȝ venus dña triplicitatis lune τ domina termini eius: aut si eēt luna in geminis in termino saturni: τ iungeret saturno: reciperet eaȝ saturnus: qꝛ est dñs triplicitatis τ termini. Cūqȝ fuerit luna vel plta in tali similitudine erit recepta. Hec est sententia Mesehalah: in receptione triplicitatis τ termini. Et si fuerit luna in hac similitudine iuncta alicui planete: τ ipse planeta iunctus fuerit domino domus in quo est luna: aut domino eius exaltationis erit luna recepta: τ si fuerit luna vacua cursu: post hec transierit ad alterum signuȝ τ iuncta fuerit domino pmi signi aut domino exaltationis eius: erit luna recepta: τ si fuerit planete iuncta qui non fuerit dominus pmi signi aut dominus exalta-

Zahel P

Claudius Ptolemy. *Tetrabiblos [Quadripartitum]* (Venice: 1519). Although it was written at the start of Christian era, Ptolemy's *Tetrabiblos,* which first appeared in this printed edition, contains a detailed account of astrological teachings. The triangular formations show the relation of different signs of the zodiac and the planets to one another.

Clavis Aſtrologiæ Elimata:

OR A

KEY to the whole ART OF ASTROLOGY New Filed and Poliſhed.

In Three PARTS.

CONTAINING

I. *An Introduction*; By which an Ordinary Capacity may Underſtand the Grounds thereof, and how to ſet a Figure upon any Occaſion: With the Schemes of the Cuſps of the Cœleſtial Houſes in Copper Plates, very uſeful in Horary Queſtions, *&c.*

II. *Select Aphoriſmes*; with Rules and Examples how to Reſolve or Judge all Lawful Queſtions Aſtrological, from a Radical Scheme Erected: Alſo Elections, and other neceſſary Precepts of Art.

III. *The Genethliacal Part*; wherein is ſhewn how to Rectifie and Calculate *Nativities*, according to *Regiomontanus*, *Argol*, and *Kepler*; with ſome Varieties in the Doctrine of Directions, Revolutions, and Profections, not before publiſhed: Alſo Tables, and all other Requiſites, both for Calculation, and Demonſtration.

To which are added the *Rudolphine Tables*, whereby the Places of the *Planets* may be Calculated for any Time, paſt, preſent, or to come.

The *SECOND EDITION*, much Enlarged and Amended.

By *HENRY COLEY*, Student in the *Mathematicks*, and *Aſtrology*.

Can'ſt thou bind the ſweet Influences of the Pleiades, *or looſe the Bands of* Orion? JOB 38. 31.

LONDON, Printed for *Benj. Tooke*, and *Tho. Sawbridge*, and are to be Sold at the *Ship* in St. *Paul's Church yard*, and at the three *Flower de Luces* in *Little Brittain*, 1676.

18 *Clavis Aſtrologiæ Elimata,*

Watry; The △ Aſpect is made from Signes of the ſame Triplicity: The *Fiery* Signes and *Airy* behold the *Earthy* and *Watry* by a □, and the contrary, *&c.*

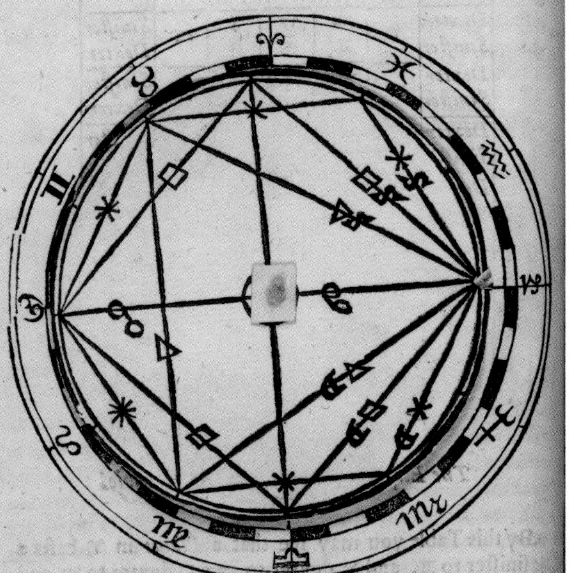

The Uſe that may be made of the *Aſpects* or *Radiations* of the *Planets*, is very conſiderable, as to the Diſcovery of ſeveral matters paſt, preſant, and to come: for ſee from what Planet, the *Moon*, Lord of the Aſcendant, or Significator, laſt ſeparated: Alſo what Planets they are in Particle Aſpect withal, and what Planet or Planets they apply unto; and ſo by conſideration of the Aſpect,

The Key to Aſtrology new Filed. 19

Aſpect, and the Planet to whom 'tis made, & what Houſe or part of the Heavens it falls in, we are enabled to Judg of what things have been paſt, what condition things are in at preſent, and laſtly, what for the future may realy be expected: And note that the power of an Aſpect is ſaid to continue twelves hours before and after the time thereof.

CHAP. III.

Of the Deſcription and Significations of the Twelve Signes of the Zodiack.

I. THe Reaſon why theſe Conſtellations of the 12 Signes are thus called by the names of ſeveral Creatures, is partly for diſtinction, and partly for that when the ☉ poſſeſſes thoſe ſeveral *Signes*, he cauſes a various alteration of the ſeaſons of the year, and makes the temperature of the Air inclinable to the Nature and Conſtitutions of thoſe ſeveral Creatures from whence they receive their Denominations; of theſe Names are many Poetical Stories. But chiefly becauſe thoſe Stars in the ſeveral Signes, do repreſent (and appear) to the Eye in Form and Figure of ſuch Creatures, as ſome are pleaſed to fancie.

But this by the way, I proceed to their ſeveral Deſcriptions, and Significations: and

I. *Of Aries.*

Aries, Is an *Equinoctial*, *Cardinal*, *Eaſterly*, and *Diurnal Signe*, of the fiery *Triplicity*, hot and dry, by

C 2 *Na-*

Henry Coley. *Clavis Astrologiae Elimata, or, a Key to the Whole Art of Astrology* (2nd edition. London: Printed for Benj. Tooke and Thos. Sawbridge, 1676).

Opposite: Omnibus works such as Coley's, which dealt with the reciprocal exchange between the Earth and the heavens, were often applied to prognostication—foretelling the future.

Above: Volvelles, or rotating paper wheels incorporated in early books and manuscripts, were used in astrology to plot zodiacal and planetary positions. (Huntington Library: RB 330362)

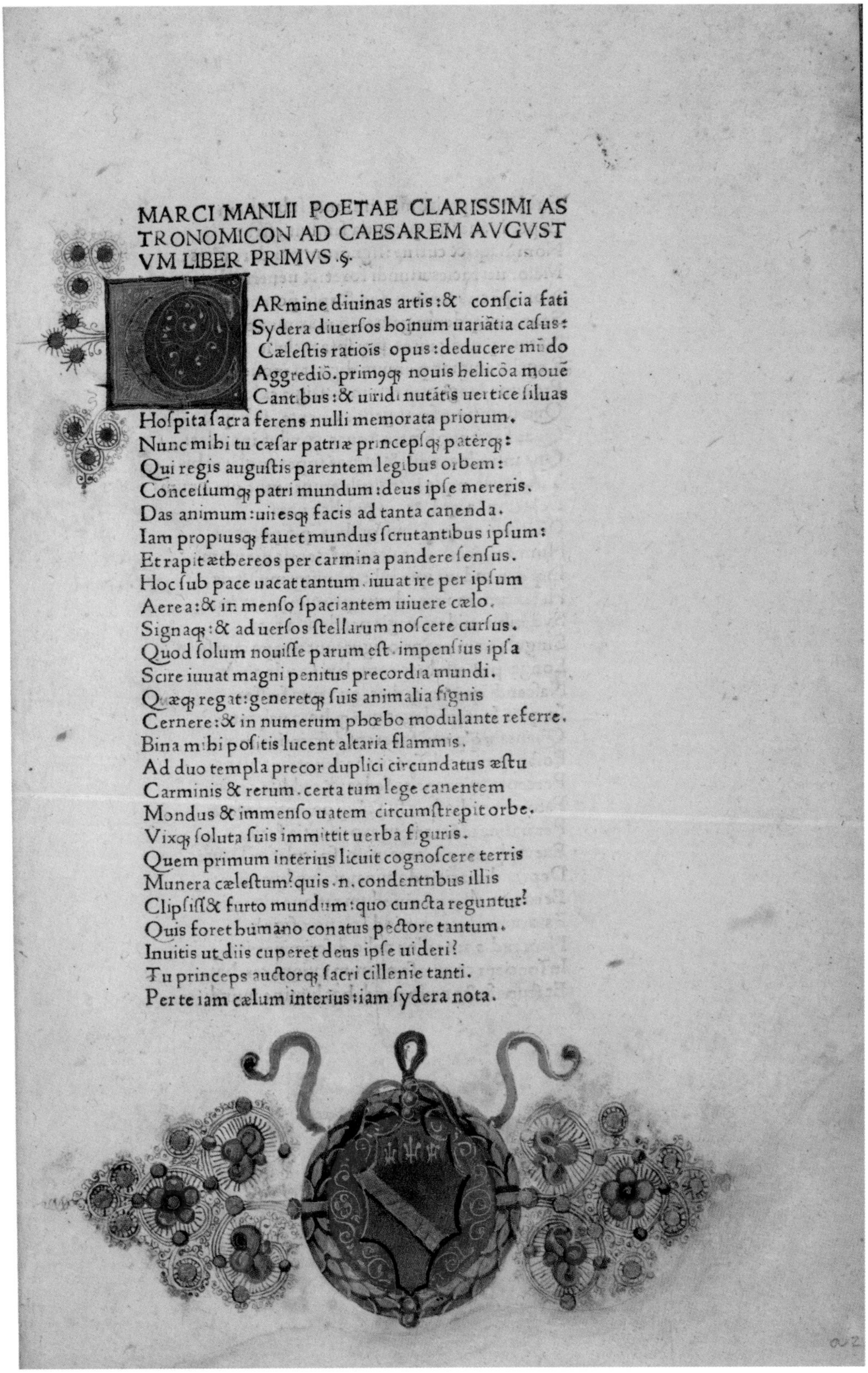

MARCI MANLII POETAE CLARISSIMI AS
TRONOMICON AD CAESAREM AVGVST
VM LIBER PRIMVS .§.

CARmine diuinas artis :& conscia fati
Sydera diuersos hoīnum uariātia casus:
Cælestis ratiōis opus :deducere mūdo
Aggrediō.primūqʒ nouis helicōa mouē
Cantibus :& uiridi nutātis uertice siluas
Hospita sacra ferens nulli memorata priorum.
Nunc mihi tu cæsar patriæ princepsqʒ paterqʒ:
Qui regis augustis parentem legibus orbem:
Concessumqʒ patri mundum :deus ipse mereris.
Das animum :uiresqʒ facis ad tanta canenda.
Iam propiusqʒ fauet mundus scrutantibus ipsum:
Et rapit ætbereos per carmina pandere sensus.
Hoc sub pace uacat tantum. iuuat ire per ipsum
Aerea :& in menso spaciantem uiuere cælo.
Signaqʒ :& aduersos stellarum noscere cursus.
Quod solum nouisse parum est. impensius ipsa
Scire iuuat magni penitus precordia mundi.
Quæqʒ regat :generetqʒ suis animalia signis
Cernere :& in numerum phœbo modulante referre.
Bina mihi positis lucent altaria flammis.
Ad duo templa precor duplici circundatus æstu
Carminis & rerum. certa tum lege canentem
Mondus & immenso uatem circumstrepit orbe.
Vixqʒ soluta suis immittit uerba figuris.
Quem primum interius licuit cognoscere terris
Munera cælestum? quis .n. condentnbus illis
Clipsisst & furto mundum :quo cuncta reguntur?
Quis foret humano conatus pectore tantum.
Inuitis ut diis cuperet deus ipse uideri?
Tu princeps auctorqʒ sacri cillenie tanti.
Per te iam cælum interius :iam sydera nota.

Marcus Manilius. *Astronomicon* (Bologne: Ugo Rugerius, 1474). Two early versions of the *Astronomicon* were printed within approximately a year: the 1473–74 version printed at the press of Regiomontanus, and this edition, from a different manuscript exemplar, a few months later. The illuminated letter is decorated with gold leaf, probably adhered with egg white. (Huntington Library: RB 103860)

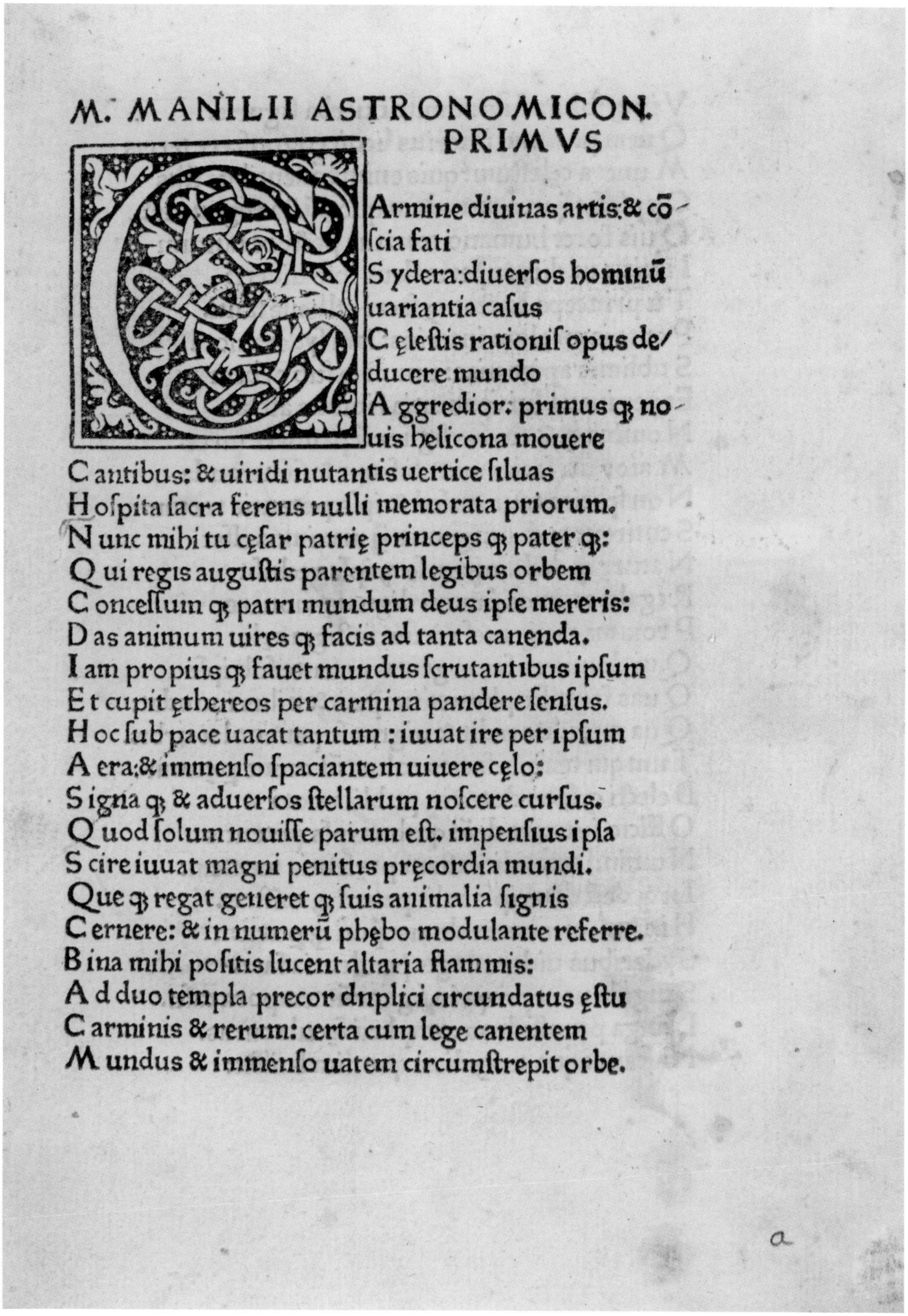

M.˙ MANILII ASTRONOMICON.
PRIMVS

CArmine diuinas artis:& cõ-
ſcia fati
Sydera:diuerſos hominũ
uariantia caſus
Cęleſtis rationiſ opus de/
ducere mundo
Aggredior: primus q; no-
uis helicona mouere
Cantibus: & uiridi nutantis uertice ſiluas
Hoſpita ſacra ferens nulli memorata priorum.
Nunc mihi tu cęſar patrię princeps q; pater q;:
Qui regis auguſtis parentem legibus orbem
Conceſſum q; patri mundum deus ipſe mereris:
Das animum uires q; facis ad tanta canenda.
Iam propius q; fauet mundus ſcrutantibus ipſum
Et cupit ętbereos per carmina pandere ſenſus.
Hoc ſub pace uacat tantum : iuuat ire per ipſum
Aera:& immenſo ſpaciantem uiuere cęlo:
Signa q; & aduerſos ſtellarum noſcere curſus.
Quod ſolum nouiſſe parum eſt. impenſius ipſa
Scire iuuat magni penitus pręcordia mundi.
Que q; regat generet q; ſuis animalia ſignis
Cernere: & in numerũ phębo modulante referre.
Bina mihi poſitis lucent altaria flammis:
Ad duo templa precor dnplici circundatus ęſtu
Carminis & rerum: certa cum lege canentem
Mundus & immenſo uatem circumſtrepit orbe.

a

Marcus Manilius. *Astronomicon* (Nuremberg: Regiomontanus, ca. 1473). The first printed book on astronomy. All of the editions from Regiomontanus's press are extremely rare: only two copies of this work have come up at auction in the past fifty years. (Huntington Library: RB 101149)

inter gradum ecliptice ascendentē & nonagesimum eius ab ascendente: uisibilis eorum cōiunctio precessit ueram. Si aūt inter eundem nonagesimū & gradum occidentem fuerit: uisibilis ueram sequetᶜ. Sed si in eodem gradu nonagesimo acciderit tunc simul uisibilis cōiunctio cum uera fiet. nullaq; diuersitas aspectus in longitudine continget. Nonagesimi⁹ nanq; gradus ecliptice ab ascendente semp est in circulo per cenith & polos zodiaci ꝓcedēte. Latitudo Lunę uisa est arcus circuli magni ꝑ polos zodiaci & locum Lunę uerū aut uisum transeuntis iuter eclipticā & circulum sibi equidistantem incedentem per locum uisum interceptus. Digiti ecliptici dicuntᶜ duodecimę diametri corporis solaris aut lunaris eclipsatę. Minuta casus i eclipsi lunari sunt minuta zodiaci quę Luna pambulat Solem supando a principio eclipsis usq; ad medium eius: si particularis fuerit: aut uniuersalis sine mora. uel a principio usq; ad initium totalis obscurationis si uniuersalis cum mora fuerit. Minuta morę dimidię sunt minuta zodiaci quę Luna Solem supando a principio totalis obscurationis usq; ad medium eius perambulat. Minuta casus i eclipsi Solari sunt miuuta quę Luna a principio eclipsis usq; ad mediū supatione

THEORICA ECLIPSIS LVNARIS.

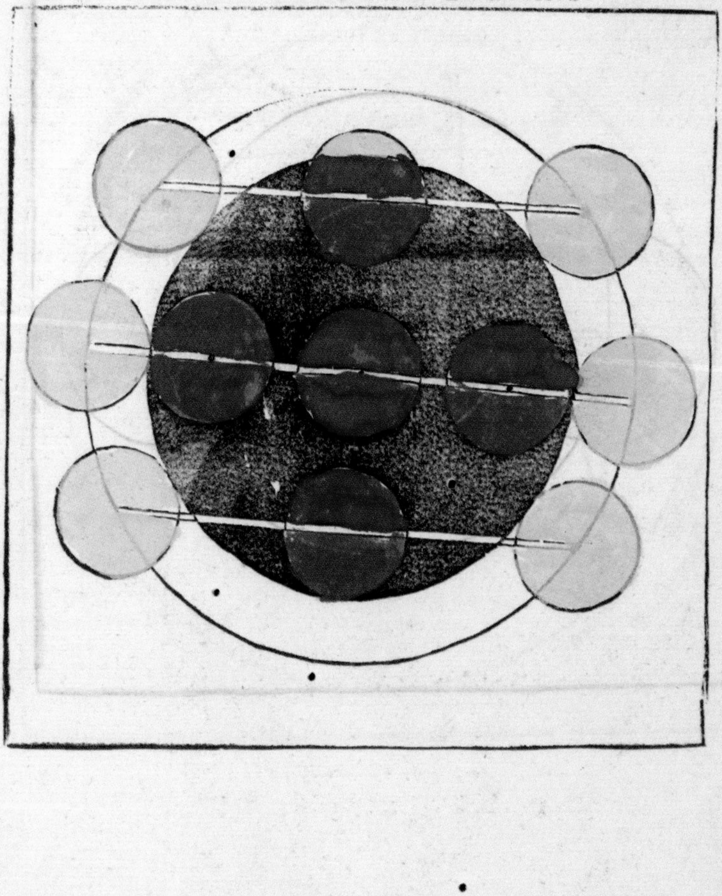

TWO | THE PTOLEMAIC UNIVERSE

Georg von Peurbach. *Theoricae novae planetarum.* (Nuremberg: Johann Mueller of Koenigsberg [Regiomontanus], about 1474.)Three possible paths of the Moon passing through Earth's shadow, illustrating Peurbach's theory of lunar ecipses. (Huntington Library: RB 104163)

WHILE THE NIGHT SKIES looked essentially the same in 1000 C.E as they do now, explanations of astronomical phenomena were completely transformed about 1500 by Nicolas Copernicus (1473–1543) who demonstrated that the Earth was not the center of the universe. Pre-Copernican attempts to understand the heavens placed people at the center of the universe on a motionless Earth, with the Sun, Moon, planets, and stars all traveling around the Earth. This line of reasoning extended back some two thousand years to the ancient Greek philosopher and astronomers. The Greeks considered astronomy to be a branch of mathematics, but they were also concerned about the physical nature of the heavens. The Greek philosopher Aristotle (384–322 B.C.), who focused on natural philosophy rather than mathematics, believed the heavens to be perfect and unchanging. Aristotle concluded that the heavens were of a different substance, called *aether*, and that things made of aether could only travel in the most perfect path —a circle. Claudius Ptolemy (ca. 100–180 C.E.), in his careful observations of the heavens, took Aristotelian conceptions further by creating a mathematically logical and precise system, based on observation, of the motions and positions of heavenly bodies.

The astronomer's primary task was to account for all the observed celestial phenomena, even the most irregular motions of the planets, by only assuming that they traveled in perfect uniform circular motion. Although it is a Christian concept of the Aristotelian universe, the basic model is well illustrated by a figure in the *Liber chronicarum* (Book of chronicles or, more commonly, Nuremberg chronicles) of 1493 by Hartmann Schedel (1440–1514). This book, a history of the world from the creation to the date of its publication, describes how the universe was created sphere by sphere. At the center of everything was the Earth, made of the spheres of the four elements: earth, water, air, and fire. These spheres were surrounded by the spheres of the Moon, Mercury, Venus, the Sun, Mars, Jupiter, and Saturn. The final spheres contained the fixed stars and the Prime Mover, the latter causing the inner spheres to move creating the daily and annual motions of the planets and stars around the fixed central Earth. Beyond the Prime Mover resided the heavenly host, the multitude of saints and angels. This model of the universe was too simple, as it did not explain all of the motions seen in the heavens.

Anyone who observed the planetary motions over a period of time knew that they moved erratically—hardly typical of objects that were supposed to move in perfect circles around a stationary Earth. Claudius Ptolemy of Alexandria took all of the available knowledge on the heavens, synthesized it, and refined it in his

Almagest (The Greatest [astronomical compilation]). This work, one of the most important in the history of astronomy, represents the culmination of Greek astronomical thought. Its influence was remarkable and pervasive up to the time of Copernicus in the sixteenth century. The *Almagest* was lost for some time to the European world during the Middle Ages, but the Arabic astronomers continued to use it. By the time of the twelfth century, the *Almagest* was translated from Arabic into Latin and was once again available to European astronomers. Before the advent of printing, works such as the *Almagest* were transmitted by being written down by hand into bound manuscripts. One such example, illustrated on page 29, was produced in southern France in 1279. Medieval manuscripts were painstakingly produced by a group of monks, each one taking care of a particular portion of the manuscript. It took several months to complete the process, which limited the spread of astronomical knowledge during the late medieval era.

Bartholomaeus Anglicus. Manuscript, *Livre des proprietes des choses* (French, first half of fifteenth century). Man holding a paddle with an image of an astrolabe, talking to two men against a striped night sky. (Huntington Library: HM 27523, f117v)

The *Almagest* took great pains to explain the irregular motions of the planets in terms of perfectly circular orbits. Ptolemy described an elaborate set of orbits upon orbits and argued that the variable motions of the planets were due to the fact that they travel in circular orbits, called *epicycles*, and that these epicycles themselves orbited around the Earth. That accounted for the varying distances of the planets from the Earth, but Ptolemy needed to explain even more erratic motions. In order to account for the apparent speeding up and slowing down of the planets, he postulated that the epicycles did not orbit the Earth, but an arbitrary point, called the *eccentric*, near the Earth. In order to eliminate even more small oddities in the planetary motions, Ptolemy came up with an original idea: the *equant*. The equant was an arbitrary point near the Earth from which the planets would have uniform circular motion, thus accounting for their unusual behavior as seen from Earth. The equant was a radical departure from the classical Greek conception of uniform circular motion, but it was consistent and helped maintain the *Almagest* as the dominant influence in astronomy until the time of Copernicus. The first printed edition of the work appeared in 1515 and many other editions followed soon thereafter.

The invention of the printing press in the 1450s caused a revolution in the transmission of knowledge. Though printed books were still confined to a small literate audience, the greater ease of their production allowed for much more information to be disseminated to readers. One of the earliest books printed that discussed the physical model of Ptolemy's universe was Theoricae novae planetarum (New planetary theory) by Georg von Peurbach (1423–1461). Peurbach's elementary textbook, published posthumously in 1474, became an influential and often reprinted work because it described how Ptolemy's universe could be physically possible. Peurbach suggested that each planet traveled as part of a set of crystalline spheres that rotated in a nonuniform manner. So by the early sixteenth century, astronomers were envisioning an extraordinary universe of planets with multiple orbits revolving around a central Earth in solid crystal spheres. *RB*

These pages from a fifteenth-century French Book of Hours represent the *Annunciation to the shepherds* (left) and the *Adoration of the Magi* (right). Even in Biblical times, stars provided a means to find directions to places, as evidenced by the Magi seeking out the baby Jesus. Before humans understood the nature and position of the earth and stars, especially bright items in the night sky could readily be granted religious signifigance as well. (Huntington Library: HM 1099)

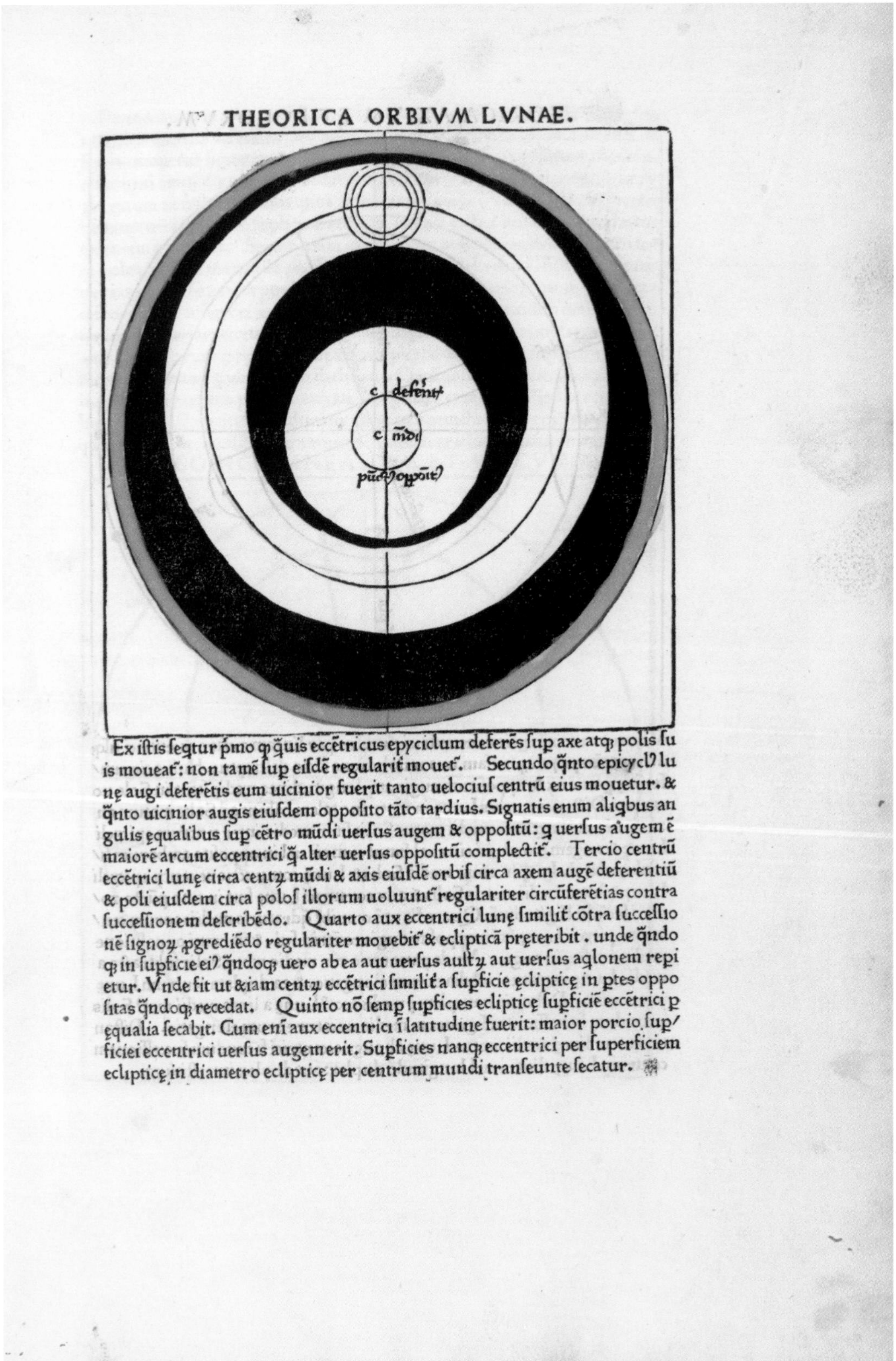

THEORICA ORBIVM LVNAE.

c defent

c mūdi

pūc9 oppōit9

Ex iſtis ſeqtur pmo q ꝙuis eccētricus epyciclum deferēs ſup axe atq polis ſuis moueat: non tamē ſup eiſdē regularit mouet. Secundo ꝙnto epicycl9 lune augi deferētis eum uicinior fuerit tanto uelociuſ centrū eius mouetur. & ꝙnto uicinior augis eiuſdem oppoſito tāto tardius. Signatis enim aliqbus angulis equalibus ſup cētro mūdi uerſus augem & oppoſitū: q uerſus augem ē maiorē arcum eccentrici ꝙ alter uerſus oppoſitū complectit. Tercio centrū eccētrici lune circa centꝝ mūdi & axis eiuſdē orbiſ circa axem augē deferentiū & poli eiuſdem circa poloſ illorum uoluunt regulariter circūferētias contra ſucceſſionem deſcribēdo. Quarto aux eccentrici lune ſimilit cōtra ſucceſſionē ſignoꝝ pgrediēdo regulariter mouebit & eclipticā preteribit. unde ꝙndoq in ſupficie ei9 ꝙndoq uero ab ea aut uerſus auſtꝝ aut uerſus aqlonem repietur. Vnde fit ut &iam centꝝ eccētrici ſimilit a ſupficie ecliptice in ptes oppoſitas ꝙndoq recedat. Quinto nō ſemp ſupficies ecliptice ſupficiē eccētrici p equalia ſecabit. Cum enī aux eccentrici ī latitudine fuerit: maior porcio ſupficiei eccentrici uerſus augem erit. Supficies nanq eccentrici per ſuperficiem ecliptice in diametro ecliptice per centrum mundi tranſeunte ſecatur.

Georg von Peurbach. *Theoricae novae planetarum*. (Nuremberg: Johann Mueller of Koenigsberg [Regiomontanus], about 1474) This very thorough work on planetary theory contains quite detailed illustrations of the solid spheres that contained the planets as they traveled around the Earth. This diagram shows the representation of the spheres carrying the Moon. (Huntington Library: RB 104163)

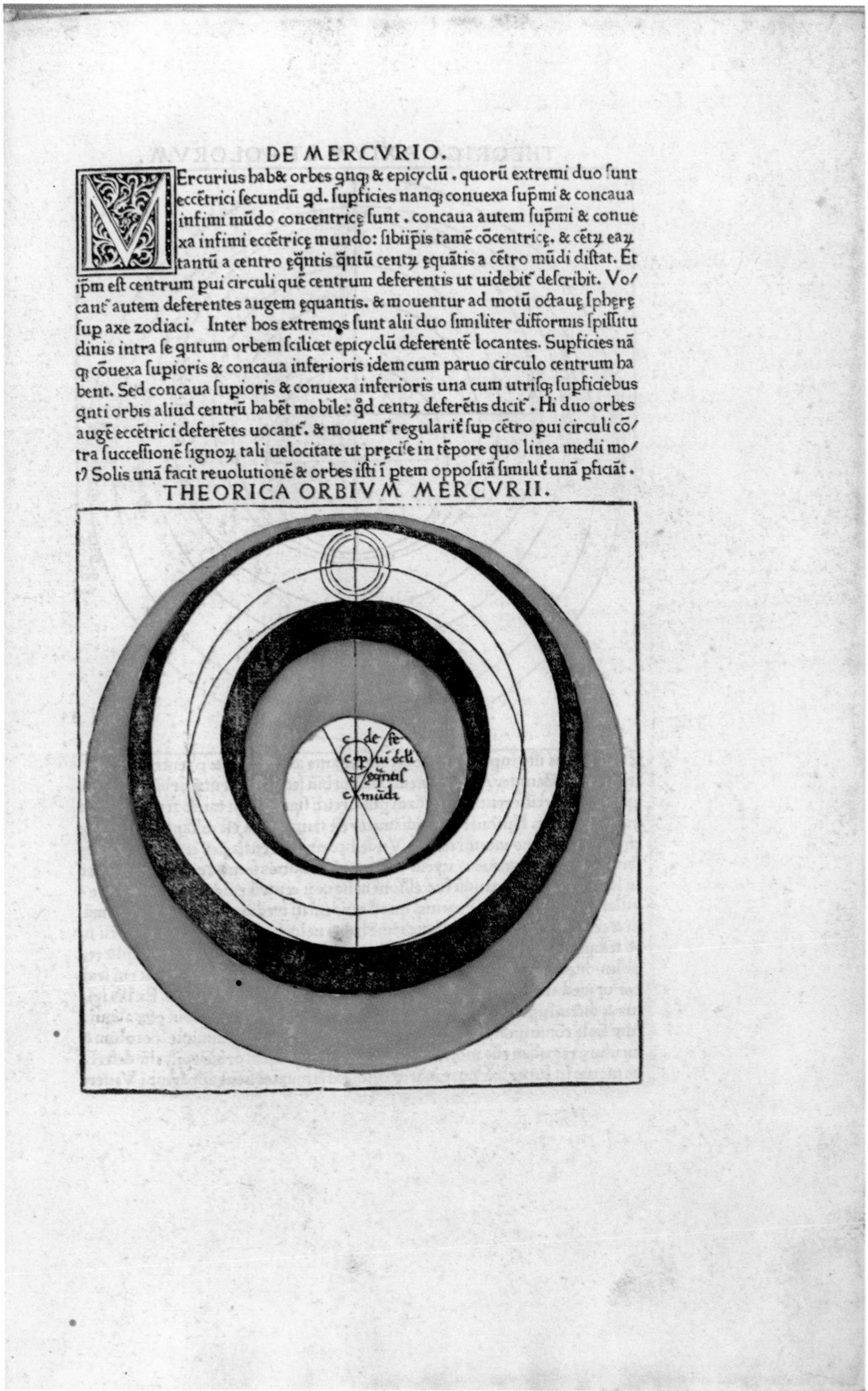

DE MERCVRIO.

Mercurius habet orbes q̃nq; & epicyclũ. quorũ extremi duo ſunt eccẽtrici ſecundũ qd. ſupficies nanq; conuexa ſupmi & concaua infimi mũdo concentricę ſunt. concaua autem ſupmi & conuexa infimi eccẽtricę mundo: ſibiipis tamẽ cõcentricę. & cẽtꝝ eaꝝ tantũ a centro ęq̃ntis q̃ntũ centꝝ ęquãtis a cẽtro mũdi diſtat. Et ipm eſt centrum pui circuli quẽ centrum deferentis ut uidebit̃ deſcribit. Vocant̃ autem deferentes augem ęquantis. & mouentur ad motũ octauę ſphęrę ſup axe zodiaci. Inter hos extremos ſunt alii duo ſimiliter difformis ſpiſſitudinis intra ſe q̃ntum orbem ſcilicet epicyclũ deferentẽ locantes. Supficies nã q; cõuexa ſupioris & concaua inferioris idem cum paruo circulo centrum habent. Sed concaua ſupioris & conuexa inferioris una cum utriſq; ſupficiebus q̃nti orbis aliud centrũ habẽt mobile: q̃d centꝝ deferẽtis dicit̃. Hi duo orbes augẽ eccẽtrici deferẽtes uocant̃. & mouent̃ regularit̃ ſup cẽtro pui circuli cõtra ſucceſſionẽ ſignoꝝ tali uelocitate ut pręciſe in tẽpore quo linea medii motꝰ Solis unã facit reuolutionẽ & orbes iſti ĩ ptem oppoſitã ſimilit̃ unã pficiãt.

THEORICA ORBIVM MERCVRII.

Georg von Peurbach. *Theoricae novae planetarum.* (Nuremberg: Johann Mueller of Koenigsberg [Regiomontanus], about 1474) The spheres carrying the planet Mercury on its epicycle are shown in this hand-colored diagram. (Huntington Library: RB 104163)

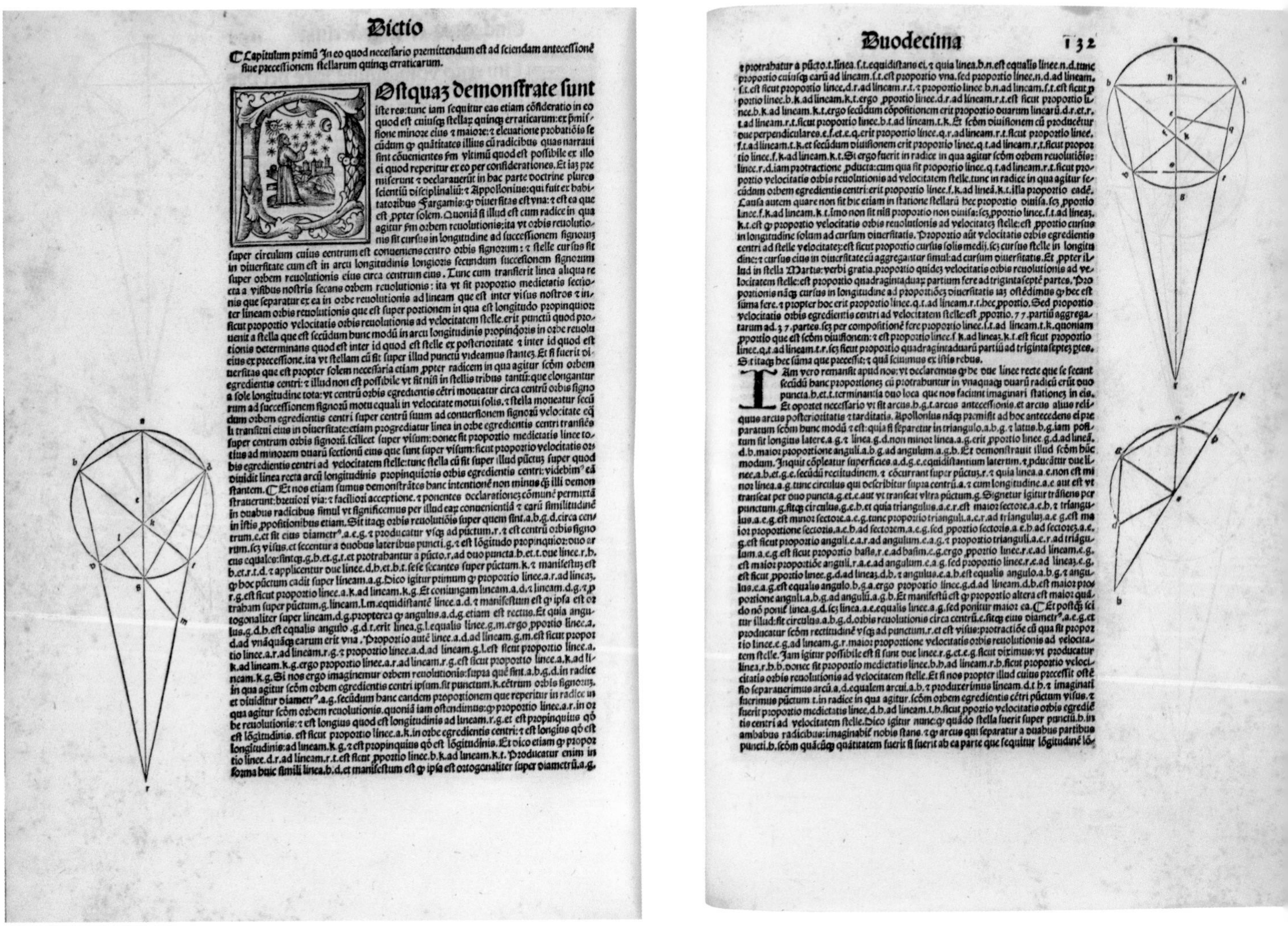

Claudius Ptolemy. *Almagest* (Venice: Petrus Liechtenstein, 1515). Ptolemy explains retrograde motion—the apparent phenomenon of a planet briefly reversing direction—in the context of a geocentric universe. (Huntington Library: RB 485907)

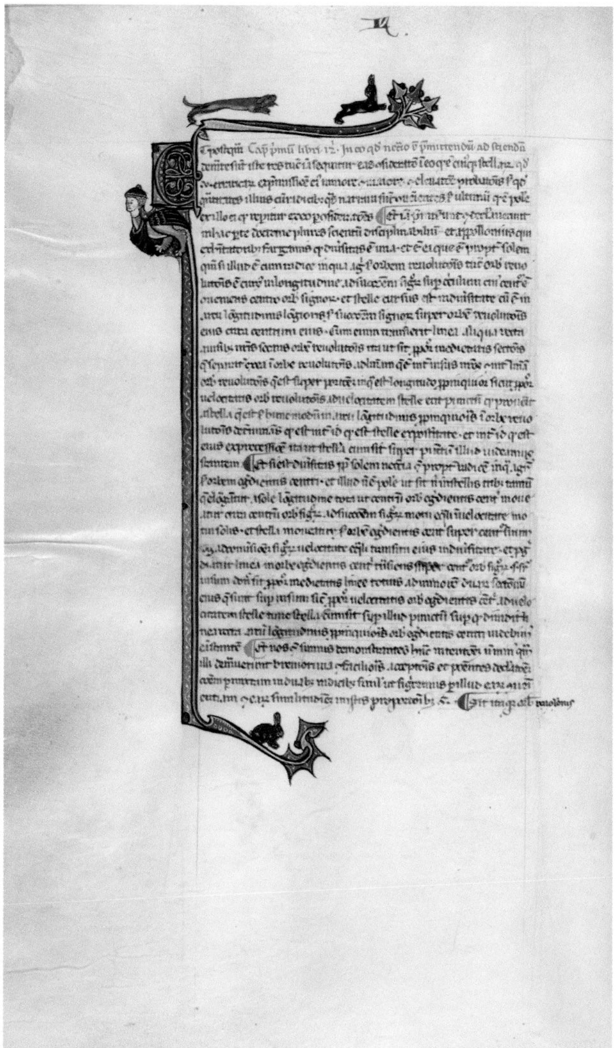

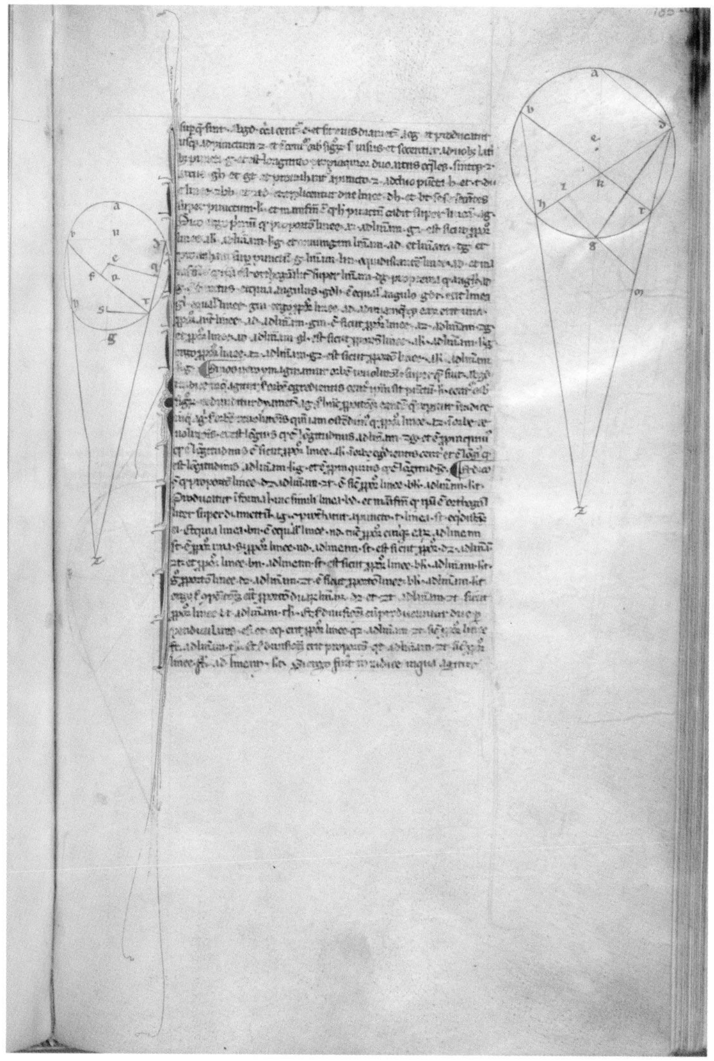

Claudius Ptolemy. Manuscript of *Almagest* (Southern France, 1279. Scribe unknown).

LEFT: This thirteenth-century manuscript on vellum, a copy of Gerard of Cremona's translation of Ptolemy's great work, was produced by several monks at a monastery in southern France. Before the printing press was developed, this was the primary medium for the communication of knowledge.

RIGHT: These geometrical diagrams from Book XII of the *Almagest* are part of Ptolemy's discussion of the motion of a planet's epicycle. (Huntington Library: HM 65)

DIALOGO
di
GALILEO GALILEI LINCEO
AL SER.mo FERD. II. GRAN. DVCA DI
TOSCANA
Stefan Della Bella

THREE

AFTER PTOLEMY

Changing Conceptions of the Universe

Nicolas Copernicus was an official of the Catholic Church in Poland, but in his spare time he pursued his true interest: astronomy. In 1496 Copernicus left the University of Cracow to study at the University of Bologna. In this new intellectual climate of Renaissance humanism, Copernicus became aware of both the scientific and classical aesthetic shortcomings of Ptolemy's universe. He posited a construct of the universe that had the Sun, rather than the Earth, as its center. Copernicus had no logical proof that the universe was heliocentric, but he could defend his views with aesthetic and rhetorical arguments that carried considerable weight in learned circles. Because he was a Catholic official, Copernicus cautiously let his views be known only among trusted friends. Eventually, Copernicus was persuaded to allow his major work, *De revolutionibus orbium coelestium* (On the revolutions of the celestial spheres), to be published in 1543 (the year of his death). While some of Copernicus's ideas were revolutionary—including the notion that the Earth orbits the Sun and rotates on its axis—he did retain such vestiges of earlier schemes such as perfectly circular orbits and epicycles. *De revolutionibus* is a landmark work in the history of science, marking a milestone in Western thought. A second edition followed the first in 1566. Approximately 500 copies of each edition were printed and over one-half of those survive to the present day. The large number of survivors attests to the high esteem in which it was held by its owners, primarily astronomers, mathematicians, and scholars. *De revolutionibus* was a highly technical work and would be incomprehensible to someone unskilled in the mathematical astronomy of the time. The copy in the Huntington Library, which is a second edition, is particularly notable in that it was formerly owned by the most famous astronomer of the twentieth century, Edwin Hubble (1889–1953), whose contributions to this science are discussed in chapter four.

Galileo Galilei. *Dialogo... di due massimi sistemi del mondo Tolemaico, e Copernicano...* (Florence: Per Gio: Batista Landini, 1632). The famous frontispiece from the *Dialogo* showing not the characters that are part of the work, but rather (from left to right) Aristotle, Ptolemy, and Copernicus. (Huntington Library: RB 297304).

Copernicus's theory of a heliocentric universe did not gain widespread acceptance upon its appearance in 1543. The authority of the scriptures, Aristotle, and Ptolemy was not to be overthrown in haste. Nevertheless, astronomers recognized the merits of Copernicus's system and some attempted to incorporate parts of it into new models. The model that gained the widest acceptance was devised by the Danish astronomer Tycho Brahe (1546–1601). Tycho's compromise involved keeping the Earth at the center of the universe with the Sun orbiting, but having the planets revolve around the Sun instead of the Earth. This unusual-looking model had the advantage of being acceptable to the theologians but also somewhat pleasing to astronomers who preferred the mathematical advantages offered by the heliocentric system. Because the orbits of some planets intersected others in the Tychonic

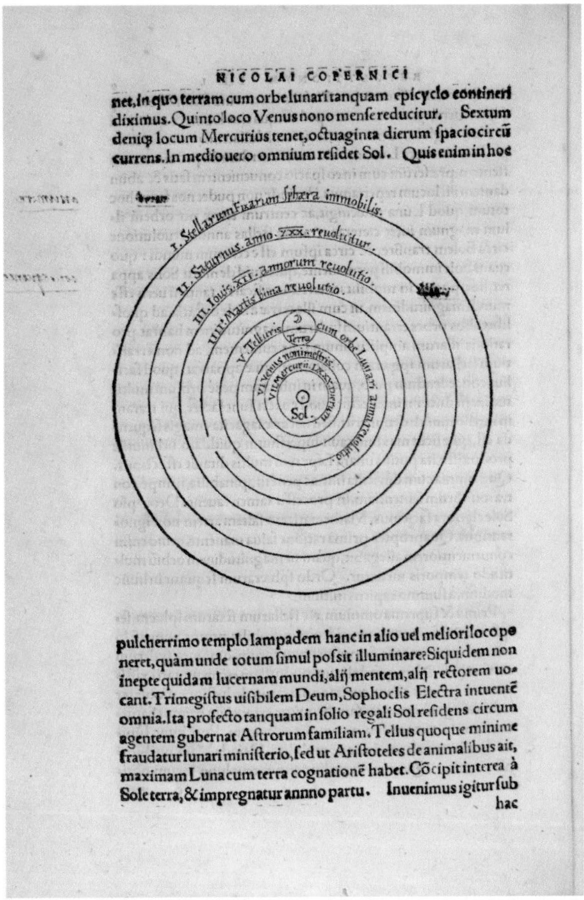

NICOLAI COPERNICI

net, In quo terram cum orbe lunari tanquam epicyclo contineri diximus. Quinto loco Venus nono menſe reducitur. Sextum deniq; locum Mercurius tenet, octuaginta dierum ſpacio circũ currens. In medio uero omnium reſidet Sol. Quis enim in hoc

pulcherrimo templo lampadem hanc in alio uel meliori loco poneret, quàm unde totum ſimul poſsit illuminare? Siquidem non inepte quidam lucernam mundi, alij mentem, alij rectorem uocant. Trimegiſtus uiſibilem Deum, Sophoclis Electra intuentẽ omnia. Ita profecto tanquam in ſolio regali Sol reſidens circum agentem gubernat Aſtrorum familiam. Tellus quoque minime fraudatur lunari miniſterio, ſed ut Ariſtoteles de animalibus ait, maximam Luna cum terra cognationẽ habet. Cõcipit interea à Sole terra, & impregnatur annno partu. Inuenimus igitur ſub

hac

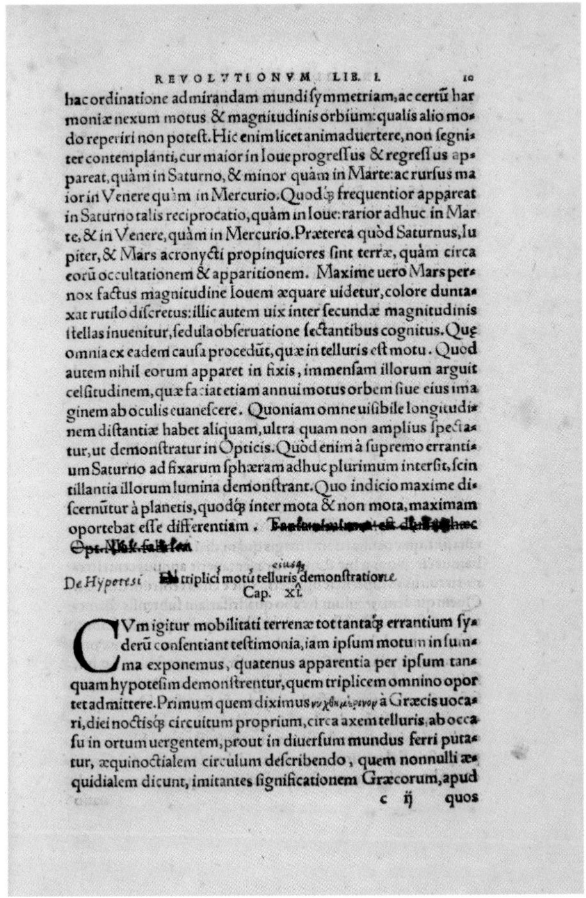

REVOLVTIONVM LIB. I. 10

hac ordinatione admirandam mundi ſymmetriam, ac certũ harmoniæ nexum motus & magnitudinis orbium: qualis alio modo reperiri non poteſt. Hic enim licet animaduertere, non ſegniter contemplanti, cur maior in Ioue progreſſus & regreſſus appareat, quàm in Saturno, & minor quàm in Marte: ac rurſus maior in Venere quàm in Mercurio. Quodq; frequentior appareat in Saturno talis reciprocatio, quàm in Ioue: rarior adhuc in Marte, & in Venere, quàm in Mercurio. Præterea quòd Saturnus, Iupiter, & Mars acronycti propinquiores ſint terræ, quàm circa eorũ occultationem & apparitionem. Maxime uero Mars pernox factus magnitudine Iouem æquare uidetur, colore duntaxat rutilo diſcretus: illic autem uix inter ſecundæ magnitudinis ſtellas inuenitur, ſedula obſeruatione ſectantibus cognitus. Quę omnia ex eadem cauſa procedũt, quæ in telluris eſt motu. Quod autem nihil eorum apparet in fixis, immenſam illorum arguit celſitudinem, quæ faciat etiam annui motus orbem ſiue eius imaginem ab oculis euaneſcere. Quoniam omne uiſibile longitudinem diſtantiæ habet aliquam, ultra quam non amplius ſpectatur, ut demonſtratur in Opticis. Quòd enim à ſupremo errantium Saturno ad fixarum ſphæram adhuc plurimum interſit, ſcintillantia illorum lumina demonſtrant. Quo indicio maxime diſcernũtur à planetis, quodq; inter mota & non mota, maximam oportebat eſſe differentiam. [illegible]

De Hypotesi ~~De~~ triplici motu telluris eiusq; demonstratione

Cap. XI.

CVm igitur mobilitati terrenæ tot tantaq; errantium ſyderũ conſentiant teſtimonia, iam ipſum motum in ſumma exponemus, quatenus apparentia per ipſum tanquam hypoteſim demonſtrentur, quem triplicem omnino oportet admittere. Primum quem diximus νυχθημερινὸν à Græcis uocari, diei noctiſq; circuitum proprium, circa axem telluris, ab occaſu in ortum uergentem, prout in diuerſum mundus ferri putatur, æquinoctialem circulum deſcribendo, quem nonnulli æquidialem dicunt, imitantes ſignificationem Græcorum, apud

c ij quos

Nicolas Copernicus. *De revolutionibus orbium coelestium* (Basel: Ex officina Henricpetrina, 1566).

LEFT: This is the famous illustration that depicts for the first time the layout of Copernicus's solar system with a central Sun and the Earth orbiting around it.

RIGHT: This page shows some of the corrections made to the text of the book. Text was crossed out in the middle of the page and the words, *"de hypotesi"* (the hypothesis) added to the phrase *"triplici motu telluris demonstratione"* (the triple motion of the earth demonstrated), in order to meet the Catholic Church's requirement that the controversial ideas of Copernicus be shown as theory and not fact. If these corrections were made by the book's owner, then it would no longer be a banned book. Copies with corrections such as these show us that their owners undoubtedly lived in or near Italy, as owners in Protestant countries and distant Catholic lands did not concern themselves with the Church's demands. (Huntington Library: RB 487000.60)

system, those who believed that the planets were carried in crystalline spheres rejected it. Tycho himself believed that the orbit of the famous comet of 1577 traveled through some of the spheres without any difficulty, proving to his satisfaction that the planetary spheres were not solid. A closer look at Tycho's universe in his posthumous *Astronomiae instauratae progymnasmata* (Introductory exercises toward a restored astronomy) of 1610 reveals that in his model the orbits of Mercury and Venus allow for their planets to exhibit phases like the Moon, something that could not happen in Ptolemy's universe but which was allowed in Copernicus's. Tycho began work on this book during his lifetime and set up a printing press at his observatory so that he could reproduce the work himself. Although he printed some of the pages, he did not complete the work, leaving his followers to complete the job in 1602. Most of these copies were purchased by a German printer who then reissued the work under his own imprint in 1610 (although using the original pages from the earlier printings).

Two of the most important supporters of Copernicus were Galileo Galilei (1564–1642) and Johannes Kepler (1571–1630). Their work in physics and planetary theory much improved the Copernican model, making it more acceptable to many astronomers. Galileo, not satisfied with the physics demanded by Aristotle's followers, set about devising a new set of mechanical principles that would better describe the motions of objects. In the course of his studies, Galileo came around to the views of Copernicus and began defending the heliocentric model against its critics. Galileo's most ardent defense came in the form of his *Dialogo...sopra i due massimi sistemi*

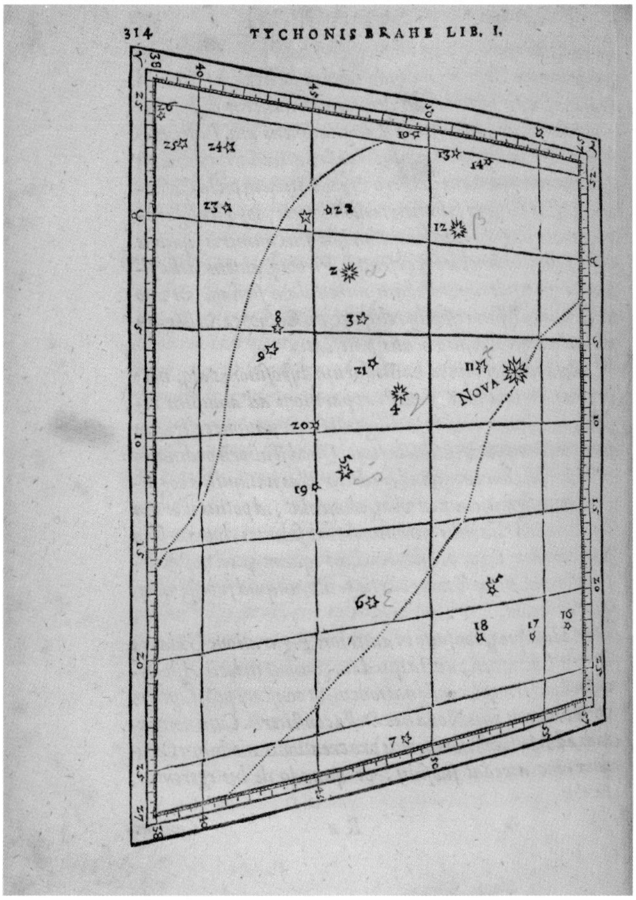

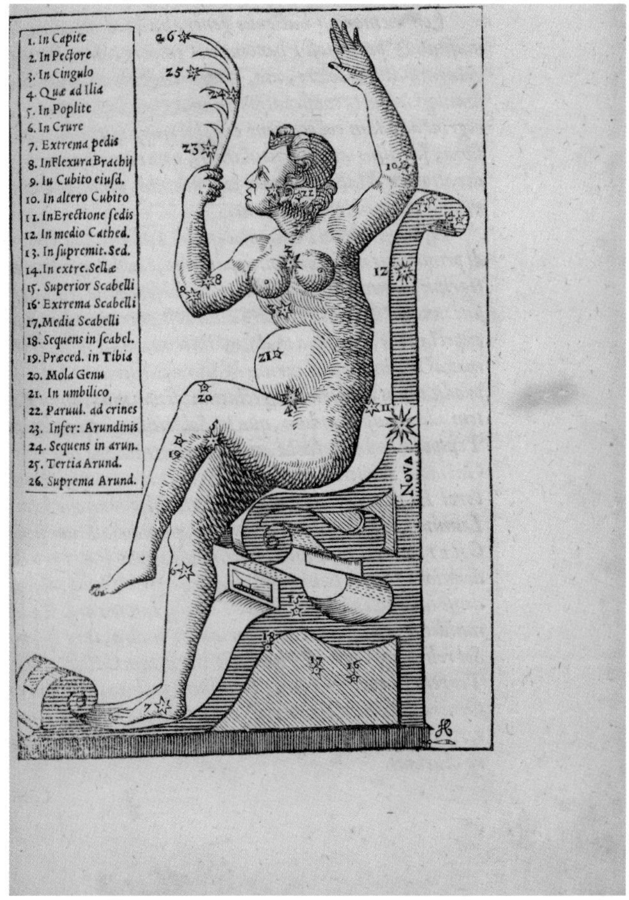

Tycho Brahe. *Astronomiae instauratae progymnasmata*... [v. 1] (Frankfurt: Godefridum Tampachium, 1610).

LEFT: This map of the sky around the constellation Cassiopeia shows the nova (new star) of 1572 at the right. Tycho's measurements indicated that this nova was beyond the Moon and thus in the heavens, which were construed to be perfect and unchanging.

RIGHT: A more illustrative map of the heavens depicting the figure of Cassiopeia with the 1572 nova located in the back of her chair at the right. (Huntington Library: RB 487000.57)

del mondo (Dialogue concerning the two chief world systems), published in 1632. This time, however, he had gone too far in supporting Copernicus over the objections of the Church: he was placed under permanent house arrest in 1633 by the Inquisition and remained in a Florentine villa until his death in 1642. Kepler, who encountered the Copernican system while studying at Tübingen in Germany, became a strong supporter of the Polish cleric by 1597. Kepler's breakthrough, the belief that the Sun provided a force that kept the planets in their orbits, provided a physical rationale for backing the heliocentric system. In 1600 Kepler became assistant to Tycho Brahe at Prague where the great Danish astronomer had taken up residence as astronomer to Rudolf II, the Holy Roman Emperor. When Tycho died in 1601, Kepler was appointed imperial mathematician and thus gained access to Tycho's valuable observational data. By 1609 Kepler announced that the Earth and planets traveled around the Sun in elliptical, not circular, orbits. Kepler's new theory of planetary motion allowed him to produce the most accurate tables of planetary positions by far, and demonstrated to astronomers that his modification to the Copernican model was the superior one.

With Aristotle and his concept of physics falling from favor, a new system was needed to explain why the planets move around the Sun. Kepler thought he had the answer but his system, which based on a solar force, did not appeal to astronomers. Galileo's mechanical physics was fine, but he did not explain the forces that caused motion. The French philosopher René Descartes (1596–1650) took it upon himself to discover the true physics of the universe. He developed a model of a mechanical

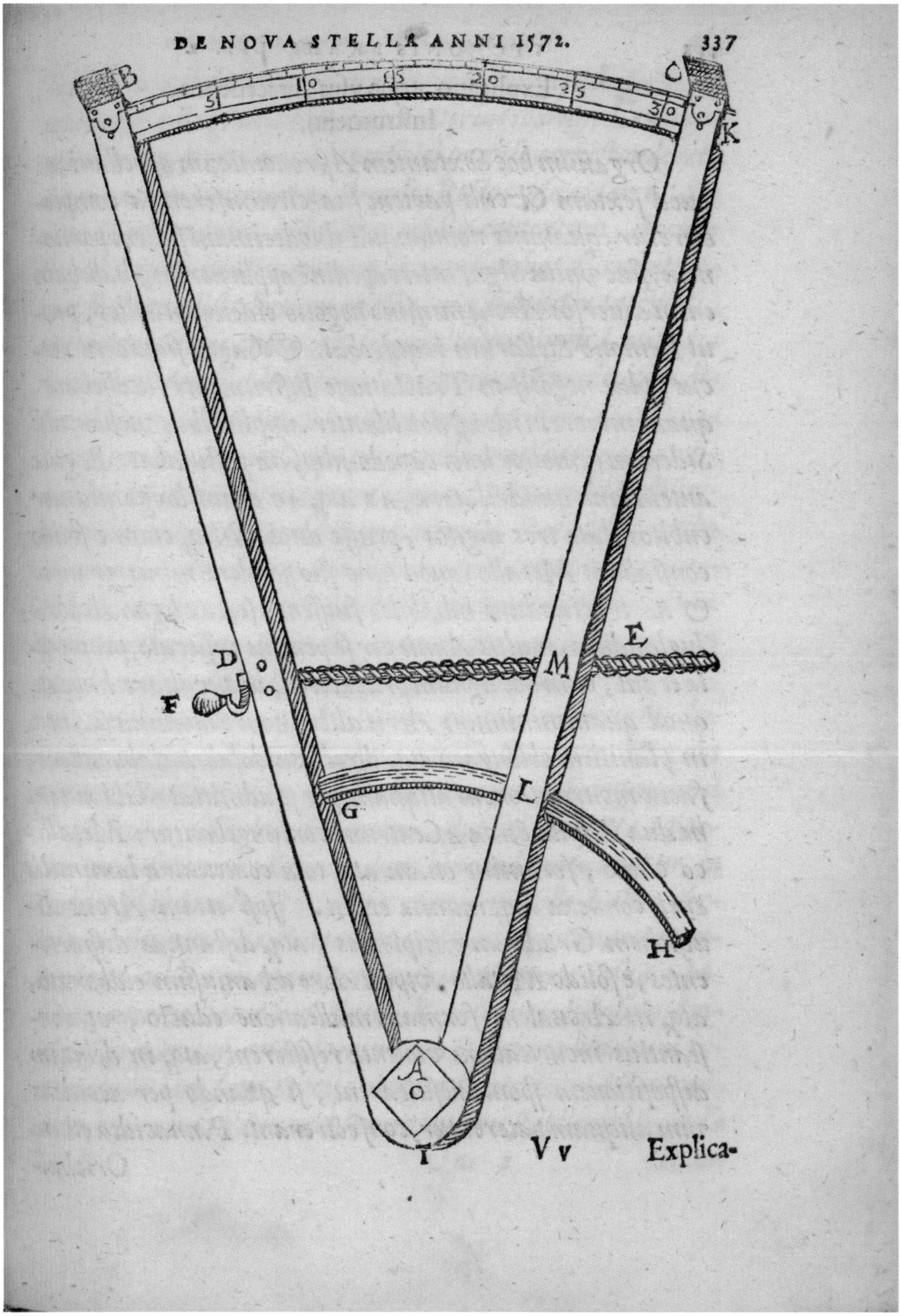

Tycho Brahe. *Astronomiae instauratae progymnasmata...* [v. 1] (Frankfurt: Godefridum Tampachium, 1610). A woodcut illustration of a large steel sextant, one of Tycho's many astronomical instruments. (Huntington Library: RB 487000.57)

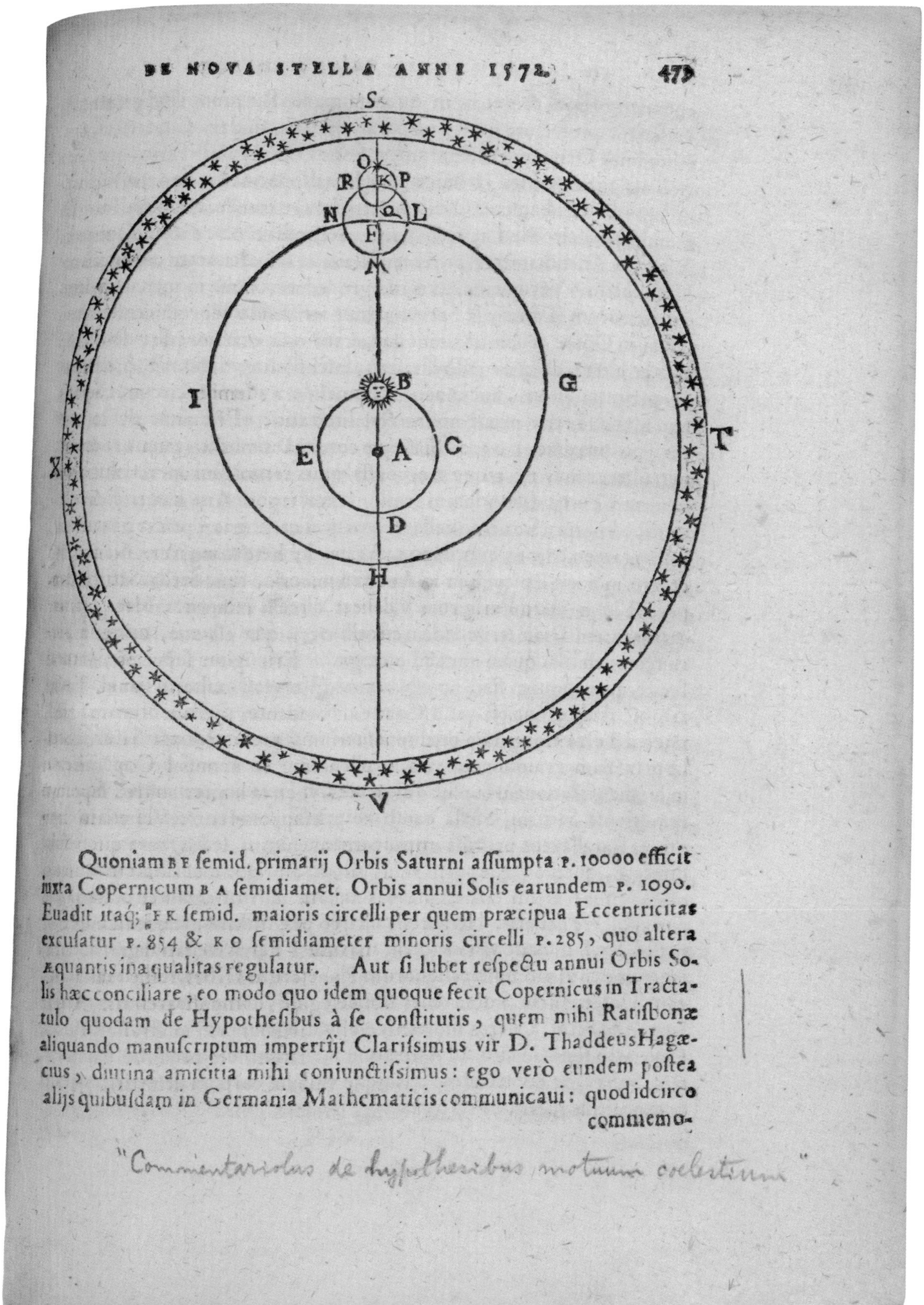
DE NOVA STELLA ANNI 1572. 479

Quoniam B F semid. primarij Orbis Saturni assumpta P. 10000 efficit iuxta Copernicum B A semidiamet. Orbis annui Solis earundem P. 1090. Euadit itaq; F K semid. maioris circelli per quem præcipua Eccentricitas excusatur P. 854 & K O semidiameter minoris circelli P. 285, quo altera æquantis inæqualitas regulatur. Aut si lubet respectu annui Orbis Solis hæc conciliare, eo modo quo idem quoque fecit Copernicus in Tractatulo quodam de Hypothesibus à se constitutis, quem mihi Ratisbonæ aliquando manuscriptum impertijt Clarissimus vir D. Thaddeus Hagæcius, diutina amicitia mihi coniunctissimus: ego verò eundem postea alijs quibúsdam in Germania Mathematicis communicaui: quod idcirco commemo-

"Commentariolus de hypothesibus motuum coelestium"

Tycho Brahe. *Astronomiae instauratae progymnasmata...* [v. 1] (Frankfurt: Godefridum Tampachium, 1610). This is Tycho's model of the solar system, an alternative to both Ptolemy and Copernicus. In it the planets orbit the Sun, which in turn orbits the Earth, so it is nominally Copernican in structure, but Earth retains its central position in the universe. (Huntington Library: RB 487000.57)

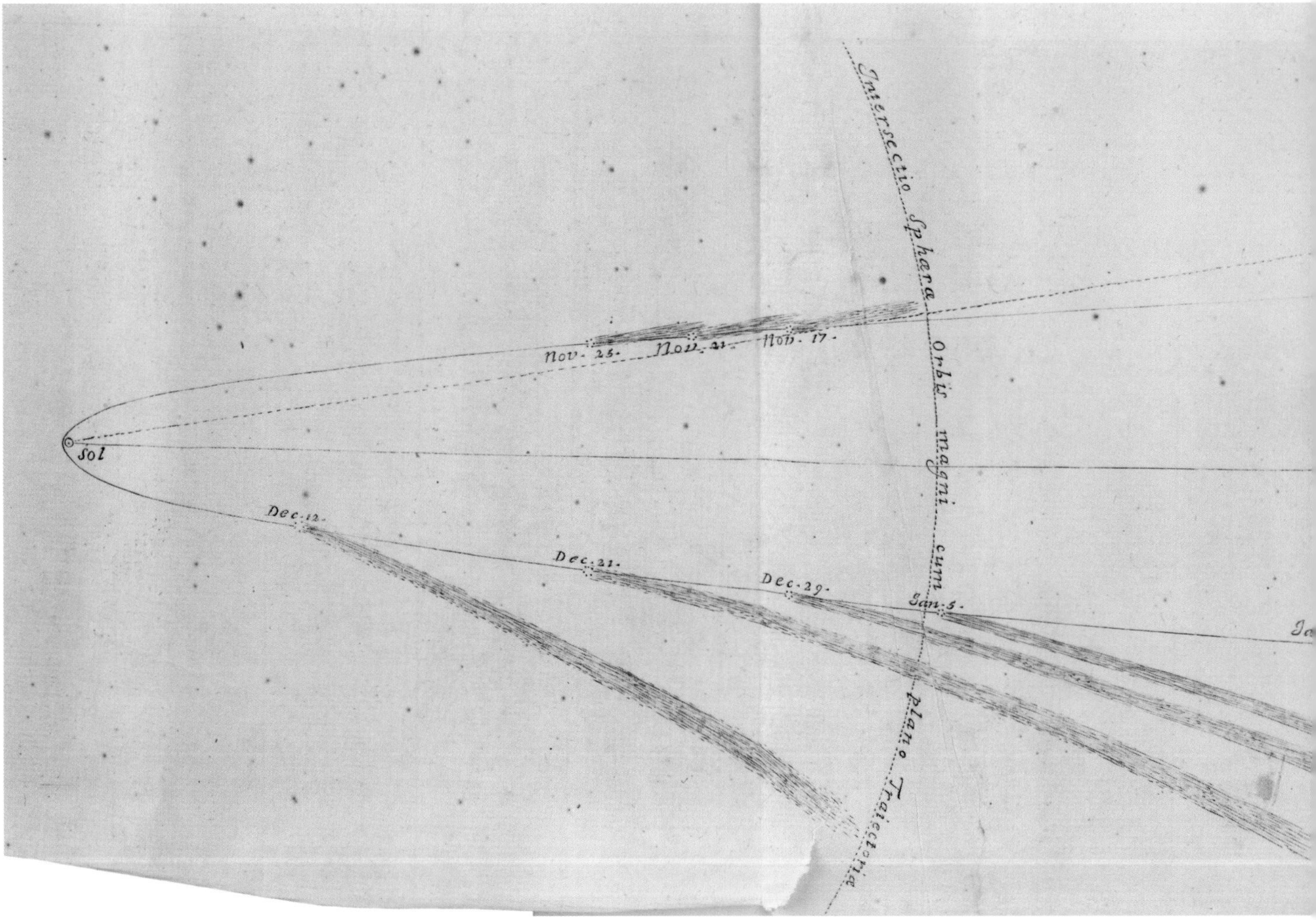

Isaac Newton. *Philosophiae naturalis principia mathematica* (London: Jussu Societatis Regi ac Typis Josephi Streater, 1687). Diagram showing the locations and appearance of the comet of 1680 (later known as Halley's Comet) in its parabolic orbit around the Sun. With this example Newton demonstrated how the trajectory of a comet can be determined from three observations. (Huntington Library: RB 42788).

universe where mind and matter were truly separate. In his *Principia philosophiae* (Principles of philosophy) of 1644, Descartes showed how matter in motion created vortices in the aether that filled the empty space in the universe. These vortices caused the motion that enabled the planets to orbit the Sun. Isaac Newton was initially impressed by the philosophers who argued for the mechanical universe and eventually developed his own system of nature, which surmounted the many difficulties that plagued the Cartesian system. The Newtonian worldview was demonstrated in his monumental 1687 work, *Philosophiae naturalis principia mathematica* (Mathematical principles of natural philosophy). Using basic geometrical methods upon which everyone could agree, Newton showed how all motion, on Earth and in the heavens, was the product of a universal force—gravity. Newton's work illustrated that there was no longer any need for an aether to transmit the forces in the universe—gravity could act at a distance through a void, long considered anathema to natural philosophers. Although Newton provided a physics that explained how the heliocentric universe worked, he could not explain exactly how gravity worked (the explanation still eludes us today). For most, however, it was sufficient that Newton's system worked. *RB*

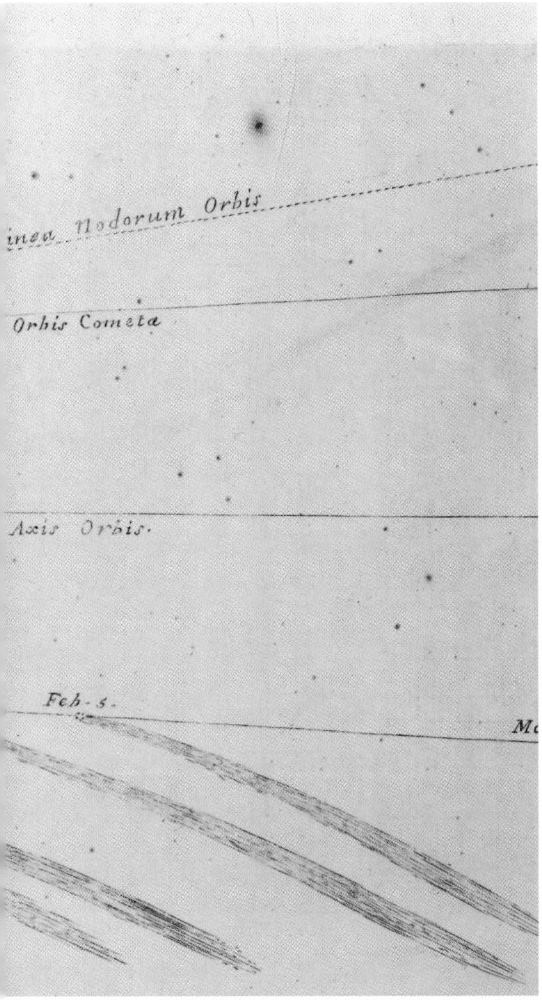

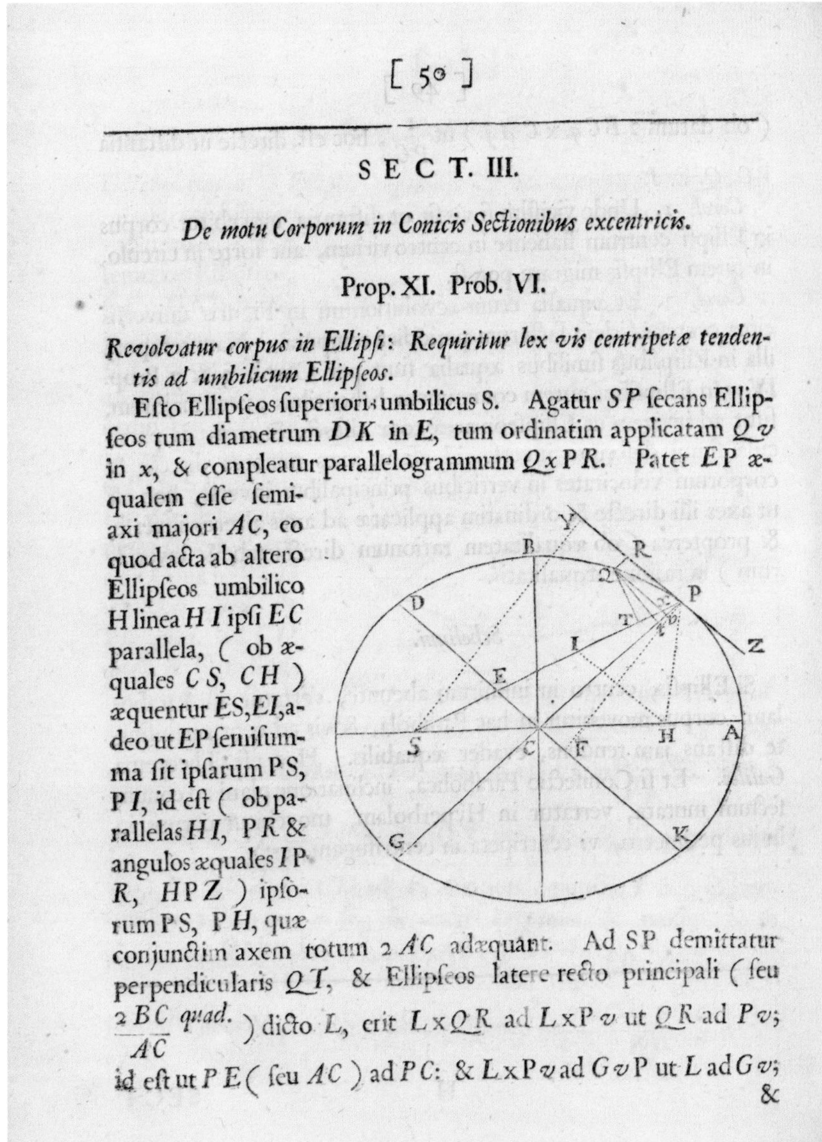

[50]

SECT. III.

De motu Corporum in Conicis Sectionibus excentricis.

Prop. XI. Prob. VI.

Revolvatur corpus in Ellipſi: Requiritur lex vis centripetæ tendentis ad umbilicum Ellipſeos.

Eſto Ellipſeos ſuperioris umbilicus S. Agatur SP ſecans Ellipſeos tum diametrum DK in E, tum ordinatim applicatam Qv in x, & compleatur parallelogrammum $QxPR$. Patet EP æqualem eſſe ſemi-axi majori AC, eo quod acta ab altero Ellipſeos umbilico H linea HI ipſi EC parallela, (ob æquales CS, CH) æquentur ES, EI, adeo ut EP ſemiſumma ſit ipſarum PS, PI, id eſt (ob parallelas HI, PR & angulos æquales IPR, HPZ) ipſorum PS, PH, quæ conjunctim axem totum $2AC$ adæquant. Ad SP demittatur perpendicularis QT, & Ellipſeos latere recto principali (ſeu $\frac{2BC\ quad.}{AC}$) dicto L, erit $L \times QR$ ad $L \times Pv$ ut QR ad Pv; id eſt ut PE (ſeu AC) ad PC: & $L \times Pv$ ad GvP ut L ad Gv; &

Isaac Newton. *Philosophiae naturalis principia mathematica* (London: Jussu Societatis Regi ac Typis Josephi Streater, 1687). Newton's famous diagram, once reproduced on British one-pound notes, illustrates the force (now called gravity) that produces the elliptical motion of the planets around the Sun. (Huntington Library: RB 42788).

Thomas Wright. *An Original Theory or New Hypothesis of the Universe*... (London: Printed for the author, 1750). Wright's drawing of a telescopic view of the Milky Way in the constellation Perseus. He used this and other images to indicate the vast number of Suns and thus the vast number of "peopled Worlds" that must exist. (Huntington Library: RB 487000.1049)

Thomas Wright. *An Original Theory or New Hypothesis of the Universe...* (London: Printed for the author, 1750). This is Wright's view of what nearby star systems must look like, with "A" (left center) being our solar system, "B" at the lower right showing the system around Sirius (the "Dog Star" in Canis Major), and "C" the system around Rigel (a bright star in Orion). (Huntington Library: RB 487000.1049)

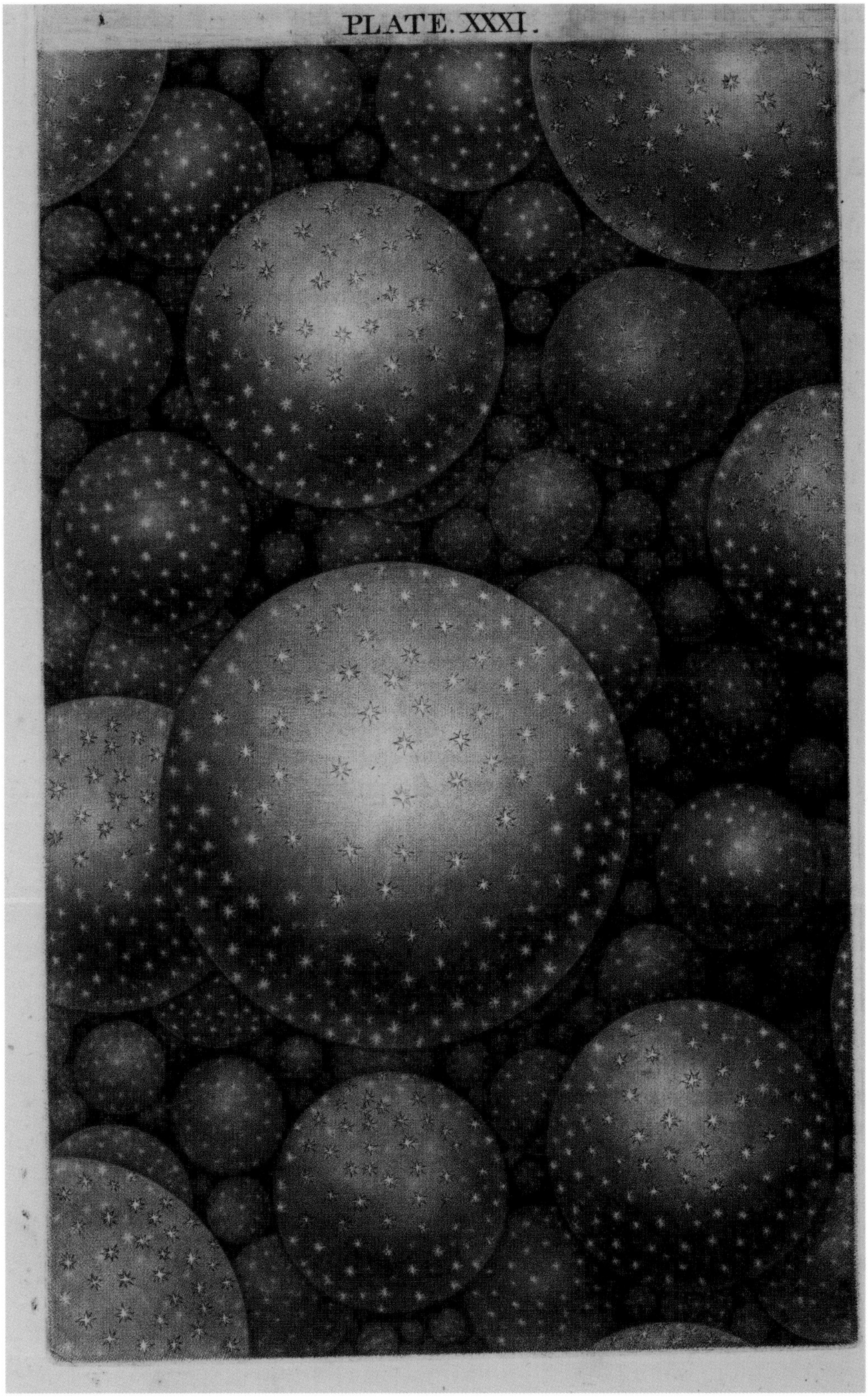

Thomas Wright. *An Original Theory or New Hypothesis of the Universe...* (London: Printed for the author, 1750). Wright's vision of the infinite number of universes that may exist, which he called "a finite view of Infinity." (Huntington Library: RB 487000.1049)

Thomas Wright. *An Original Theory or New Hypothesis of the Universe...* (London: Printed for the author, 1750). A cross-section of the universes illustrated in Wright's work. The eye in each universe symbolizes the author's belief that their creator is all-seeing, "Object of that incomprehensible Being, which alone and in himself comprehends and constitutes supreme Perfection." (Huntington Library: RB 487000.1049)

Jean Blaeu. *Le grand atlas, ou cosmographie Blaviane...* (Amsterdam: Jean Blaeu, 1663). Representation of one of Tycho Brahe's large instruments—an astronomical sextant for measuring the distance between celestial objects. (Huntington Library: RB 74701)

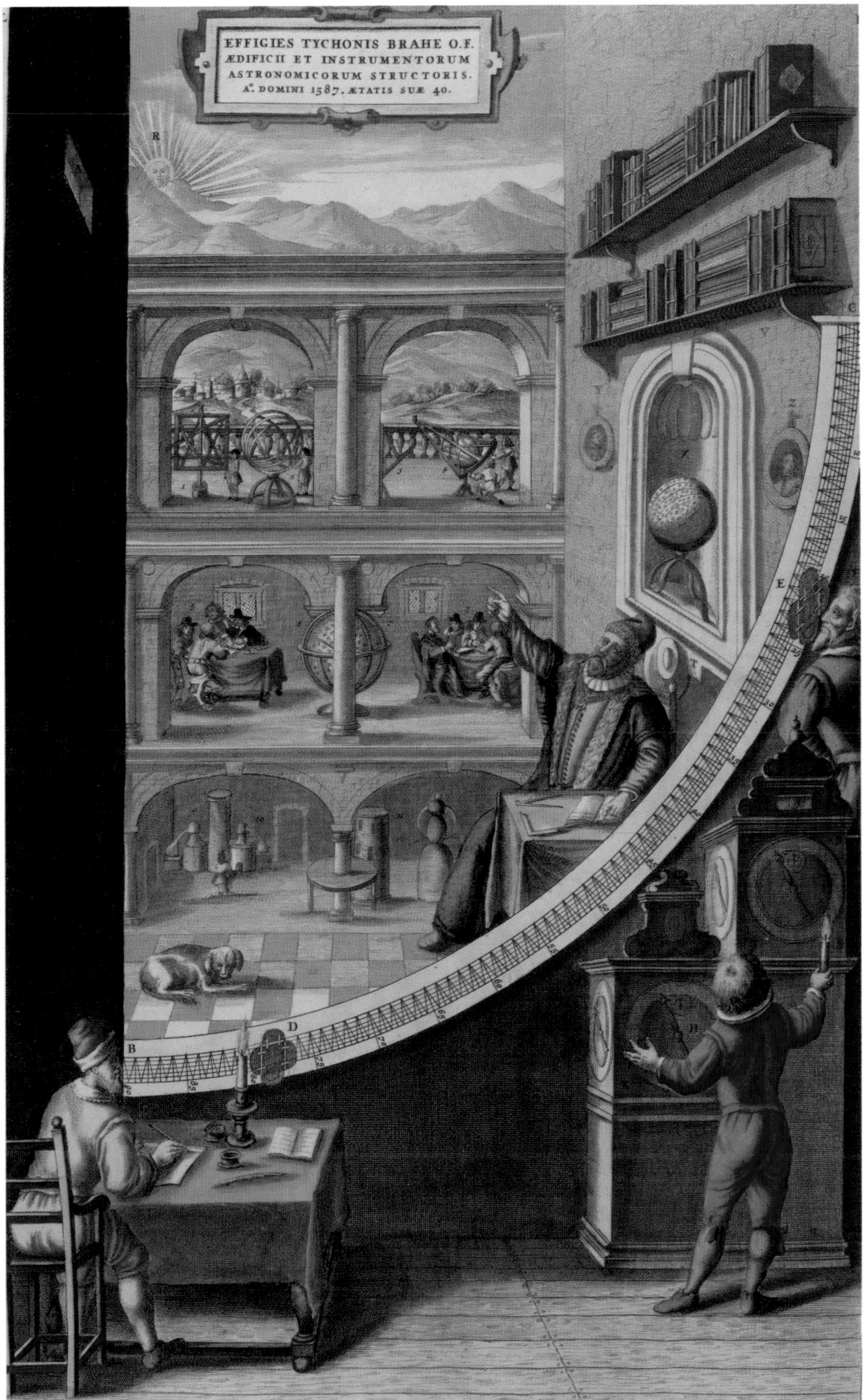

Jean Blaeu. *Le grand atlas, ou cosmographie Blaviane...* (Amsterdam: Jean Blaeu, 1663). This great mural quadrant measured over six feet from the front sight (in the opening of the wall at the upper left) to the rear sight, or pinnule, which could be moved along the arc. The quadrant was used to measure the height of celestial bodies. At lower right are two clocks that Tycho described as having "the greatest possible accuracy." (Huntington Library: RB 74701)

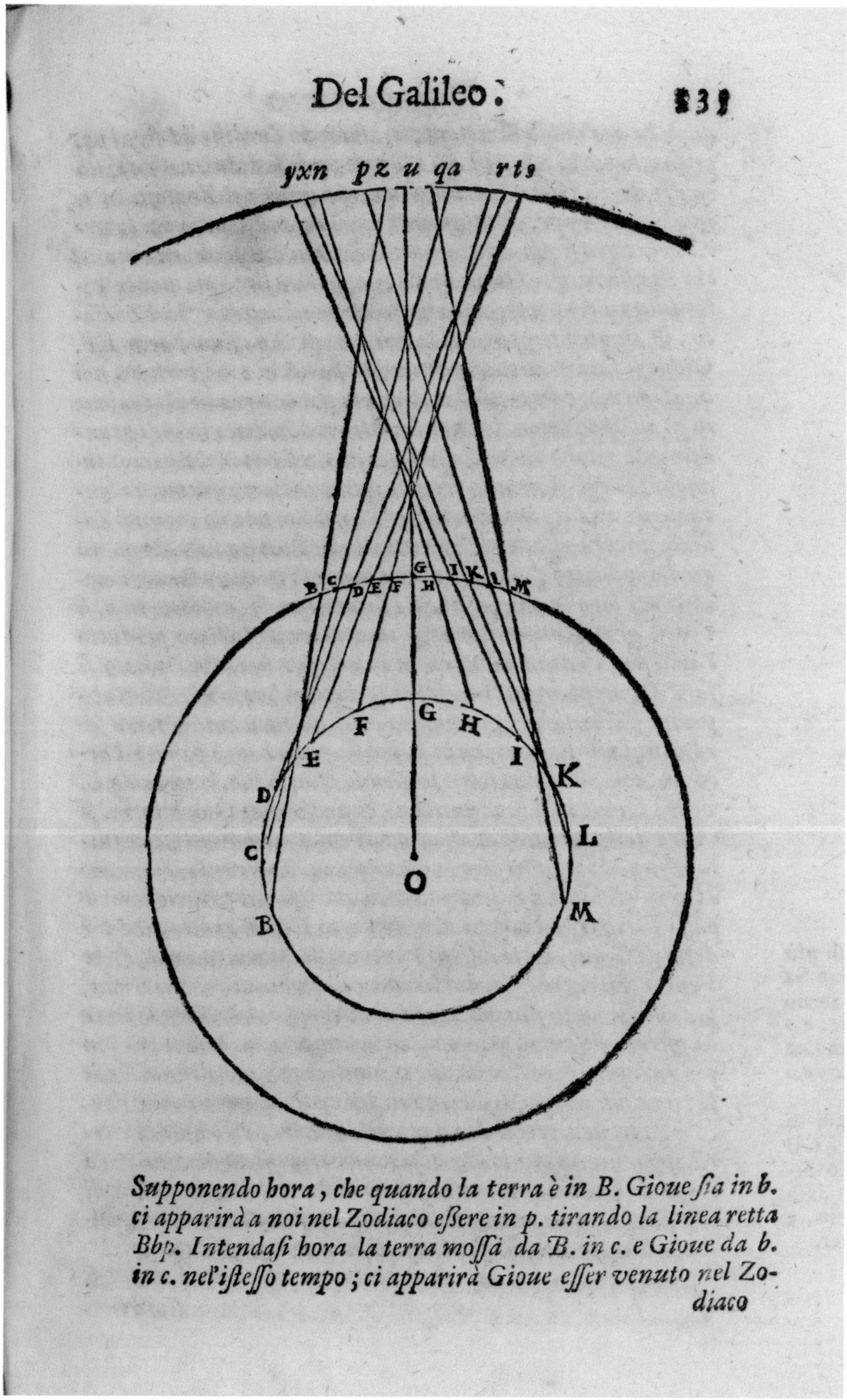

Del Galileo. 131

Supponendo hora, che quando la terra è in B. Gioue sia in b. ci apparirà a noi nel Zodiaco essere in p. tirando la linea retta Bbp. Intendasi hora la terra mossa da B. in c. e Gioue da b. in c. nel'istesso tempo; ci apparirà Gioue esser venuto nel Zodiaco

Galileo Galilei. *Dialogo... di due massimi sistemi del mondo Tolemaico, e Copernicano...* (Florence: Per Gio: Batista Landini, 1632). Diagram showing how the irregular retrograde motion of Jupiter can be explained as a consequence of the Earth's motion around the Sun (at O). As the Earth moves in its orbit (inner circle) it catches up with and passes Jupiter (in the outer circle). From Earth, Jupiter appears to move forward, stop, move backward, stop, and then move forward again. (Huntington Library: RB 297304).

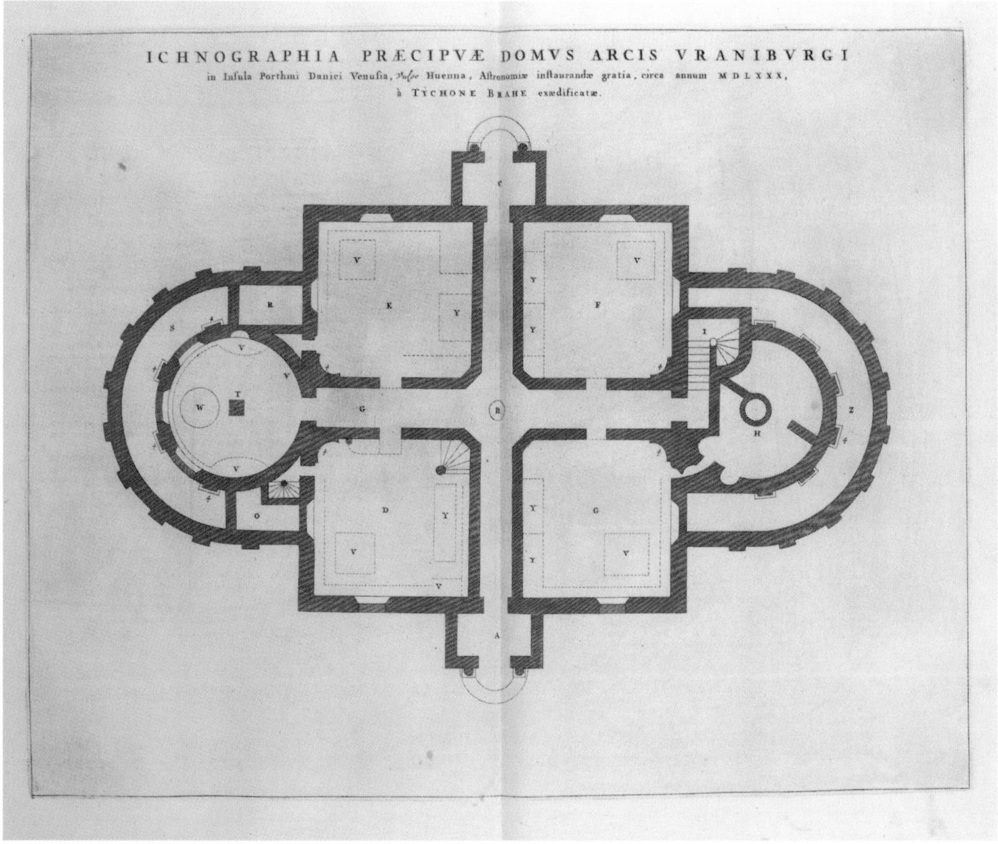

ICHNOGRAPHIA PRÆCIPVÆ DOMVS ARCIS VRANIBVRGI
in Insula Porthmi Danici Venusia, Vulgo Huenna, Astronomiæ instaurandæ gratia, circa annum MDLXXX,
à TYCHONE BRAHE exædificatæ.

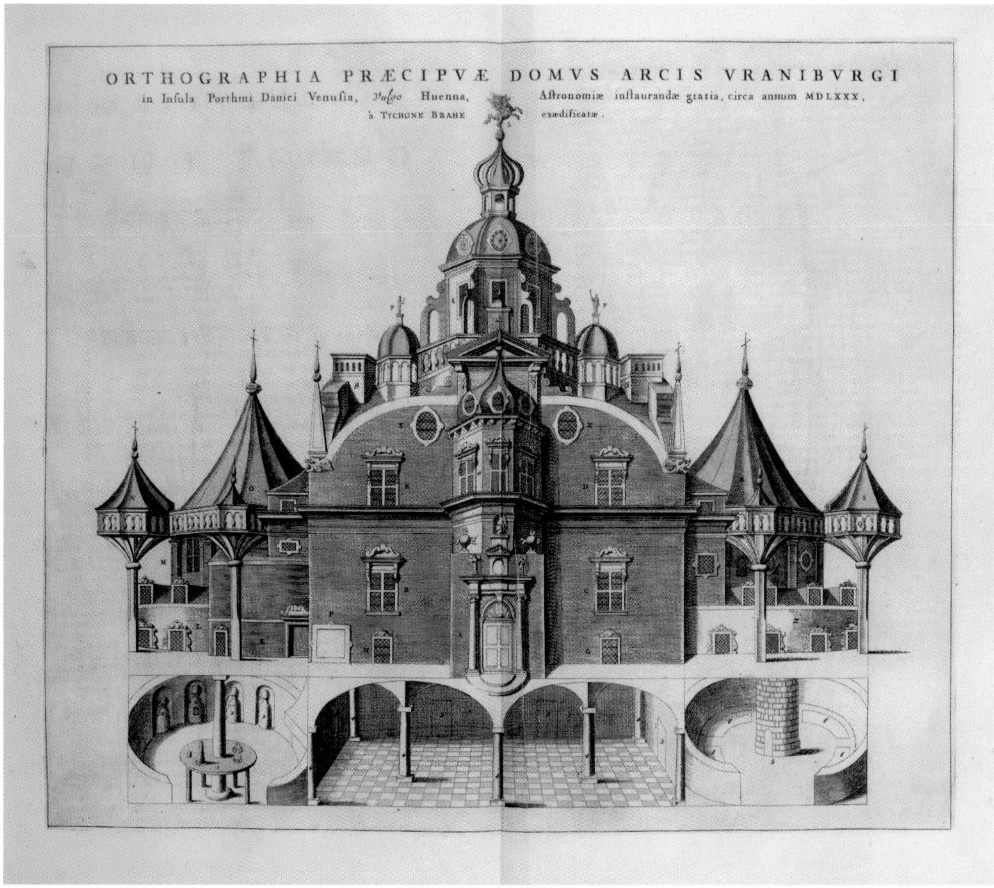

Jean Blaeu. *Le grand atlas, ou cosmographie Blaviane...* (Amsterdam: Jean Blaeu, 1663).

TOP: The instruments Tycho used for astronomical measurements were exceptionally accurate, and many were very large. His great brass globe, on which he marked the positions of approximately a thousand stars, was located in his library, in the center of the circular room on the left.

BOTTOM: Tycho Brahe's palace, the Castle of Uraniborg, was completed in 1586 on the island of Hven, in the sound between Denmark and Sweden. Many of Tycho's instruments were housed under the sliding conical roofs on the left and right. (Huntington Library: RB 74701)

STELLÆBURGUM five OBSERVATORIU
IN INSULA HVÆNA, EXTRA ARCEM URANI.
L
D
A
B
E
N
Amftelæ

Jean Blaeu. *Le grand atlas, ou cosmographie Blaviane...* (Amsterdam: Jean Blaeu, 1663). Stjerneborg, or "Star Castle," Tycho Brahe's second observatory, was also located on Hven. Most of Stjerneborg's instruments, including some of Tycho's finest, were installed below ground to protect them from the wind. (Huntington Library: RB 74701)

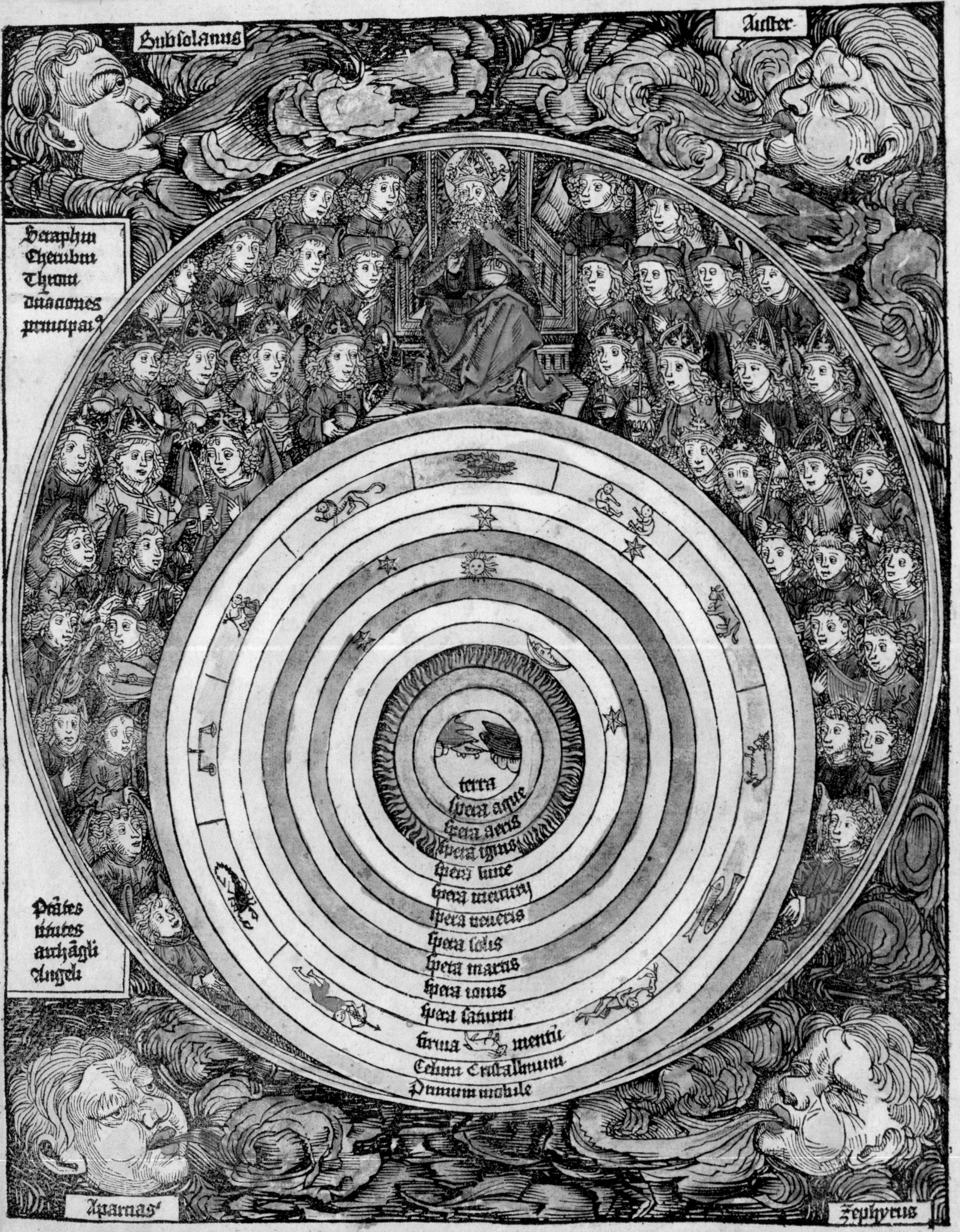

Subsolanus
Auster
Seraphin
Cherubin
Throni
dñaciones
principat9
Ptãtes
ṽtutes
archãgli
Angeli
terra
spera aque
spera aeris
spera ignis
spera lune
spera mercurij
spera veneris
spera solis
spera martis
spera iovis
spera saturni
firma mentũ
Celum Cristallinum
Primum mobile
Zephyrus

FOUR COSMOLOGY AND THE INFINITE UNIVERSE

THE CLASSICAL CONCEPTION OF THE UNIVERSE, as promulgated by Aristotle and Ptolemy, was that it did not extend infinitely in all directions. The universe had a limited extent, which was quite small compared to modern notions of cosmic scale. Campanus of Novara (d. 1296) estimated that the universe extended out to the sphere of fixed stars, some seventy million miles from the central Earth. There were a few who argued otherwise. Giordano Bruno (1548–1600) said that the universe was infinite and contained countless suns with planets orbiting them, and these views may have contributed to his being burned at the stake. By 1700 astronomers were able to get a better idea of the larger scale of the solar system and hence, the universe. The Earth was thought to be some eighty million miles from the Sun, and the stars were no longer located in a thin shell but were rather scattered about the heavens with the nearest ones some 160 million miles distant. The stars were believed to be suns just like our own and each one as far from the other as the nearest ones were to the Sun. Still, there was no way with the instruments at hand that astronomers could directly determine the distance to stars; they were just too far away.

When Copernicus placed the Earth at the center of the universe, he also had the Earth rotate on its axis every twenty-four hours to account for the daily motion of the stars and planets in the sky. There was no longer any need for the stars to be confined to a moving sphere to take them around the Earth in a day. But this implication did not sink in too deep and many astronomers continued to envision a distant and finite region of stars. One person who proved to be the exception to this finite way of thinking was Thomas Digges (1546–1595). Digges was the first to introduce the Copernican system to England in an appendix to his father Leonard's 1576 work, *Prognostication everlasting*. The appendix, titled *A perfit description of the caelestiall orbes*, contained a diagram that clearly depicted the stars scattered about the heavens outside the orbit of Saturn and, as he described it, "This orbe of starres fixed infinitely up extendeth hit self in altitude sphericallye..." But Digges still retains the old dualism of the perfect heavens and the corruptible Earth even though the latter was now placed in the heavens, orbiting the Sun.

In the eighteenth century, people studying the structure and the evolution of the universe, or cosmology, felt that the ultimate truths about the universe were beyond the realm of science. Still, that did not prevent people from speculating and philosophizing on the structure of the universe using facts ascertained from observational data. One of the best starting points for understanding the arrangement of stars in the heavens was the appearance of the Milky Way: a thin band of

OPPOSITE
Hartmann Schedel. *Liber chronicarum* (Nuremberg: Antonius Koberger, 1493). In this image depicting the seventh day of creation, God rests in his throne, surrounded by the nine orders of angels. The earth is located at the center of the universe, ringed by the perfect cosmos. (Huntington Library: RB 95858)

BELOW
Albert Einstein at Carnegie Observatories offices on Santa Barbara Street, Pasadena, late January or early February 1931. (Huntington Library: Carnegie Institution of Washington Collection, COPC 2807).

faint stars extending around in a circle at an angle to the path of the Sun and planets in the sky (the ecliptic). Thomas Wright of Durham (1711–1786) developed a model for the heavens based on moral-theological principles. In his beautifully illustrated 1750 work, *An original theory or new hypothesis of the universe*, Wright argued that the Milky Way's appearance could be due to the fact that either we are in the midst of a thin shell of stars or the Milky Way is in itself a flat ring of stars. According to Wright, though, the universe was not infinite and would not evolve from its current state. Wright's work was highly influential on Immanuel Kant (1724–1804) who produced an elaborate model for the universe in which the Milky Way was an immense disk of stars with our solar system somewhere near the center. Unlike Wright's universe, however, Kant's evolved from an early primitive form and was theoretically infinite in extent. Kant's model was not particularly influential; Johann Heinrich Lambert (1728–1777) published a similar theory around the same time, one presumably produced with no knowledge of Kant's system. The two systems were quite similar although Lambert's universe, like Wright's, was basically stable and finite.

With the twentieth-century confirmation of the existence of distant galaxies (discussed in chapter 17), the size of the universe grew dramatically. Studies of the arrangement and distribution of galaxies became part of the growing field of cosmology. When the astronomer Vesto M. Slipher (1875–1969) measured the spectra of light from several spiral nebulae in 1914, almost all of them showed evidence that they were moving away from us (spectra demonstrate evidence of motion away from or toward us through the Doppler effect—a shift of the spectra toward the red end indicates motion away from us and the object is said to have a redshift). After a number of astronomers examined this phenomenon, Edwin Hubble and his colleague Milton Humason (1891–1972) measured more than forty spiral nebulae (now being called galaxies) and found practically all of them to have redshifts. Hubble announced in his classic paper, "A relation between distance and radial velocity among extra-galactic nebulae," of 1929 that not only were most galaxies moving away from us, but that their speed was directly proportional to their distance from us (Hubble's Law). In a few years, more studies caused astronomers to realize that the universe was expanding. Since every galaxy was moving away from all the others, no one could say where the center of the universe was. The rate of expansion is measurable, and Hubble's first estimates suggested that all of the material in the universe was aggregated together at a time about one billion years ago.

BOTTOM LEFT
Walther Mayer, Albert Einstein, Walter Adams, and Edwin Hubble on the catwalk of the 100-inch telescope dome at Mount Wilson, January 1931. (Huntington Library: Carnegie Institution of Washington Collection, COPC 2794)

BOTTOM RIGHT
Walt Disney, Edwin Hubble, and Sir Julian Huxley at Disney Studio on Hyperion Avenue in Los Angeles. (Huntington Library: HUB 1054)

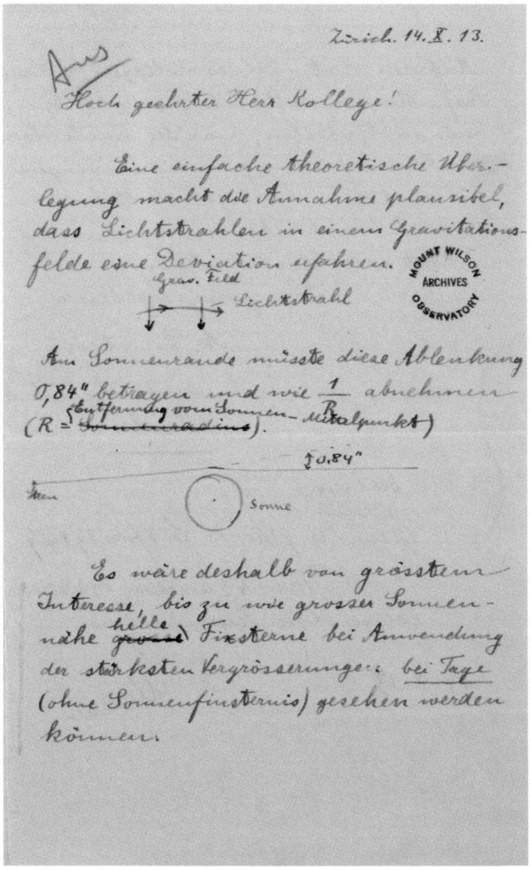

Zürich. 14. X. 13.

Hoch geehrter Herr Kollege!

Eine einfache theoretische Überlegung macht die Annahme plausibel, dass Lichtstrahlen in einem Gravitationsfelde eine Deviation erfahren.

Am Sonnenrande müsste diese Ablenkung 0,84" betragen und wie $\frac{1}{R}$ abnehmen (R = ~~Sonnenradius~~ Entfernung vom Sonnen-Mittelpunkt).

Es wäre deshalb von grösstem Interesse, bis zu wie grosser Sonnennähe ~~grosse~~ helle Fixsterne bei Anwendung der stärksten Vergrösserungen bei Tage (ohne Sonnenfinsternis) gesehen werden können.

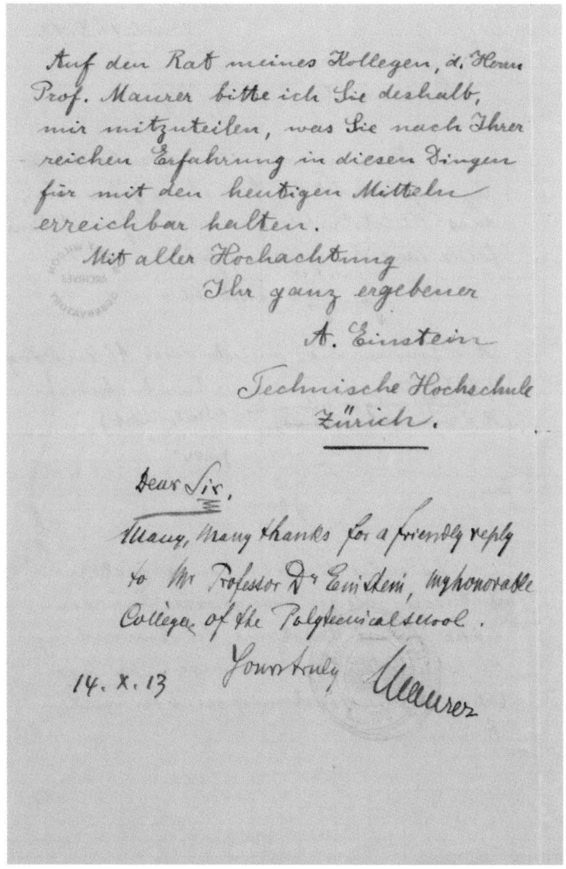

Auf den Rat meines Kollegen, d. Herrn Prof. Maurer bitte ich Sie deshalb, mir mitzuteilen, was Sie nach Ihrer reichen Erfahrung in diesen Dingen für mit den heutigen Mitteln erreichbar halten.

Mit aller Hochachtung
Ihr ganz ergebener
A. Einstein
Technische Hochschule
Zürich.

Dear Sir,

Many, many thanks for a friendly reply to Mr Professor Dr Einstein, my honorable College of the Polytechnical school.

14. X. 13 Yours truly Maurer

Letter, Albert Einstein to George Ellery Hale, 14 October 1913.

Top left: Einstein's general theory of relativity implied that light could be bent by a gravitational field. In this diagram Einstein explains to Hale his prediction of how light from a distant star could be deflected as it passed near the massive Sun.

Top right: The second page, with his signature, of Einstein's letter to Hale. (Huntington Library: Hale Papers, Mount Wilson Directors' Files, Box 154).

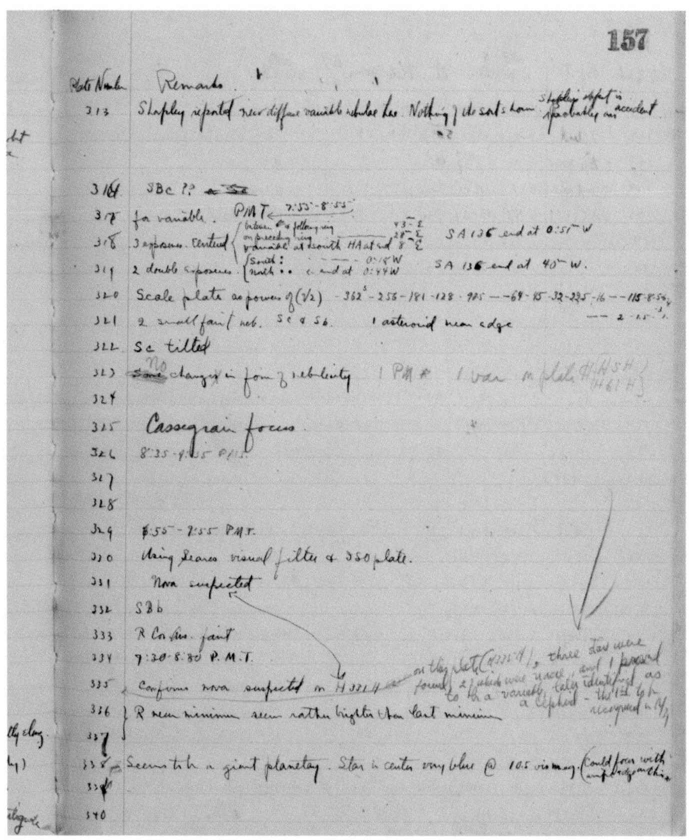

157

Plate Number — Remarks

318 3 exposures. Centered

320 Scale plate as powers of (√2)

322 Sc tilted

323 No changes in form of nebulosity

325 Cassegrain focus

329 8:55 – 9:55 P.M.T.

330 Using Seares visual filter & 350 plate.

331 Nova suspected

332 SBb

334 7:30 – 8:30 P.M.T.

335 Confirms nova suspected on H331H

336 R near minimum seen rather brighter than last minimum

338 Seems to be a giant planetary. Star in center very blue

340

RIGHT

Edwin P. Hubble. Logbook with Hubble's observations through the 100-inch Hooker Telescope, Mount Wilson. The arrow pointing to Hubble's note recording a nova in the spiral nebula of Andromeda (M31) indicates that he grasped the importance of this discovery. (Huntington Library: HUB 1098)

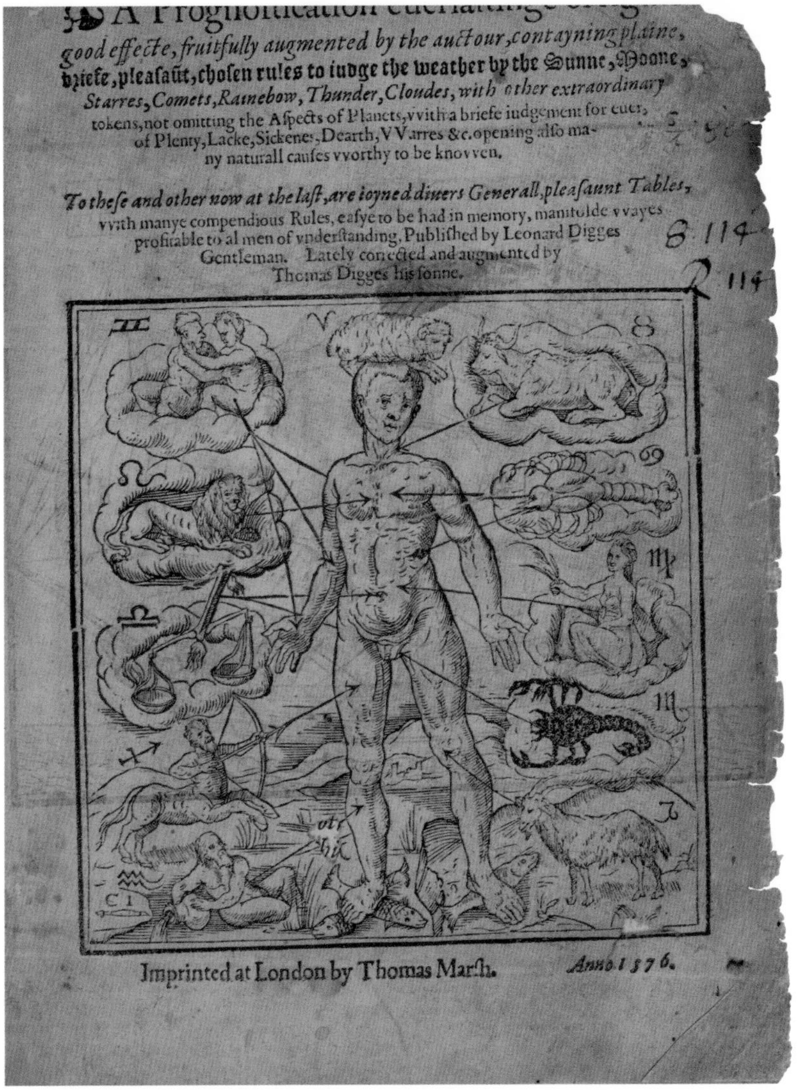

RIGHT AND OPPOSITE

Thomas Digges. *Prognostication everlasting: A perfit description of the caelestiall orbes* (London: Imprinted ... by Thomas Marshe ..., 1576). This work contained Digges's model of a Copernican universe, as illustrated at far right, showing the stars scattered out toward infinity for the first time. This illustration at top right shows which parts of the human body were controlled by the different signs of the zodiac. (Huntington Library: RB 53921)

Since Hubble and Humason's resarch, the finest minds, including Albert Einstein (1879–1955), have been working to understand the age and dimensions of the universe. In 1915 Einstein completed his general theory of relativity, a major accomplishment that took his earlier theory, one that concentrated on space-time effects on electromagnetism, and expanded it to encompass all physical phenomena, including gravitation. One of the consequences of the general theory is that a light ray would be deflected by the gravitational field of a body it passes. Einstein wrote to George Ellery Hale (1868–1938), the leading American solar astronomer, in 1913 asking him whether, in his experience, someone could detect the offset location of a star that happened to be just to the side of the Sun during normal daylight hours. Hale responded that this could not be done with instruments available at the time and that one would have to wait for a solar eclipse to view a star next to the Sun. World War I helped prevent any major eclipse expeditions from being mounted. It was not until 1919 that a British expedition led by Arthur S. Eddington (1882–1944) took photographs which demonstrated that starlight is deflected by the gravitational field of the Sun, thus providing proof for Einstein's theory. Even with similar research that allows us to improve our model of the universe, however, we are still far from a true understanding of the cosmos. *RB*

A perfit description of the Cælestiall Orbes,

according to the most auncient doctrine of the Pythagoreans. &c.

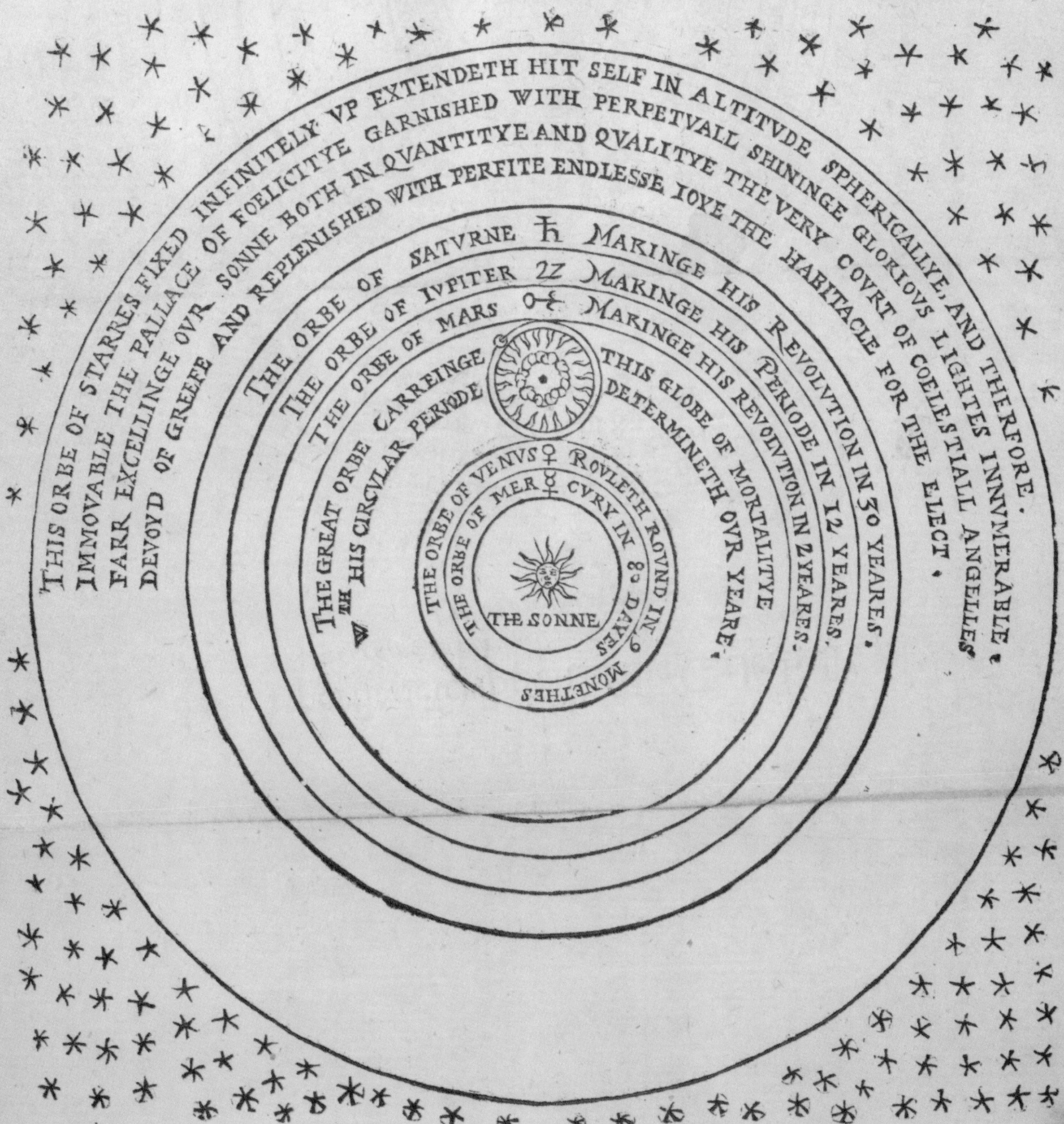

THIS ORBE OF STARRES FIXED INFINITELY VP EXTENDETH HIT SELF IN ALTITVDE SPHERICALLYE, AND THERFORE.
IMMOVABLE THE PALLACE OF FOELICITYE GARNISHED WITH PERPETVALL SHININGE GLORIOVS LIGHTES INNVMERABLE.
FARR EXCELLINGE OVR SONNE BOTH IN QVANTITYE AND QVALITYE THE VERY COVRT OF COELESTIALL ANGELLES.
DEVOYD OF GREEFE AND REPLENISHED WITH PERFITE ENDLESSE IOYE THE HABITACLE FOR THE ELECT.
THE ORBE OF SATVRNE ♄ MAKINGE HIS REVOLVTION IN 30 YEARES.
THE ORBE OF IVPITER ♃ MAKINGE HIS PERIODE IN 12 YEARES.
THE ORBE OF MARS ♂ MAKINGE HIS REVOLVTION IN 2 YEARES.
THE GREAT ORBE CARREINGE THIS GLOBE OF MORTALITYE
HIS CIRCVLAR PERIODE DETERMINETH OVR YEARE.
THE ORBE OF VENVS ♀ ROVLETH ROVND IN 9 MONETHES
THE ORBE OF MER ☿ CVRY IN 80 DAYES
THE SONNE

M. .1. A PER-

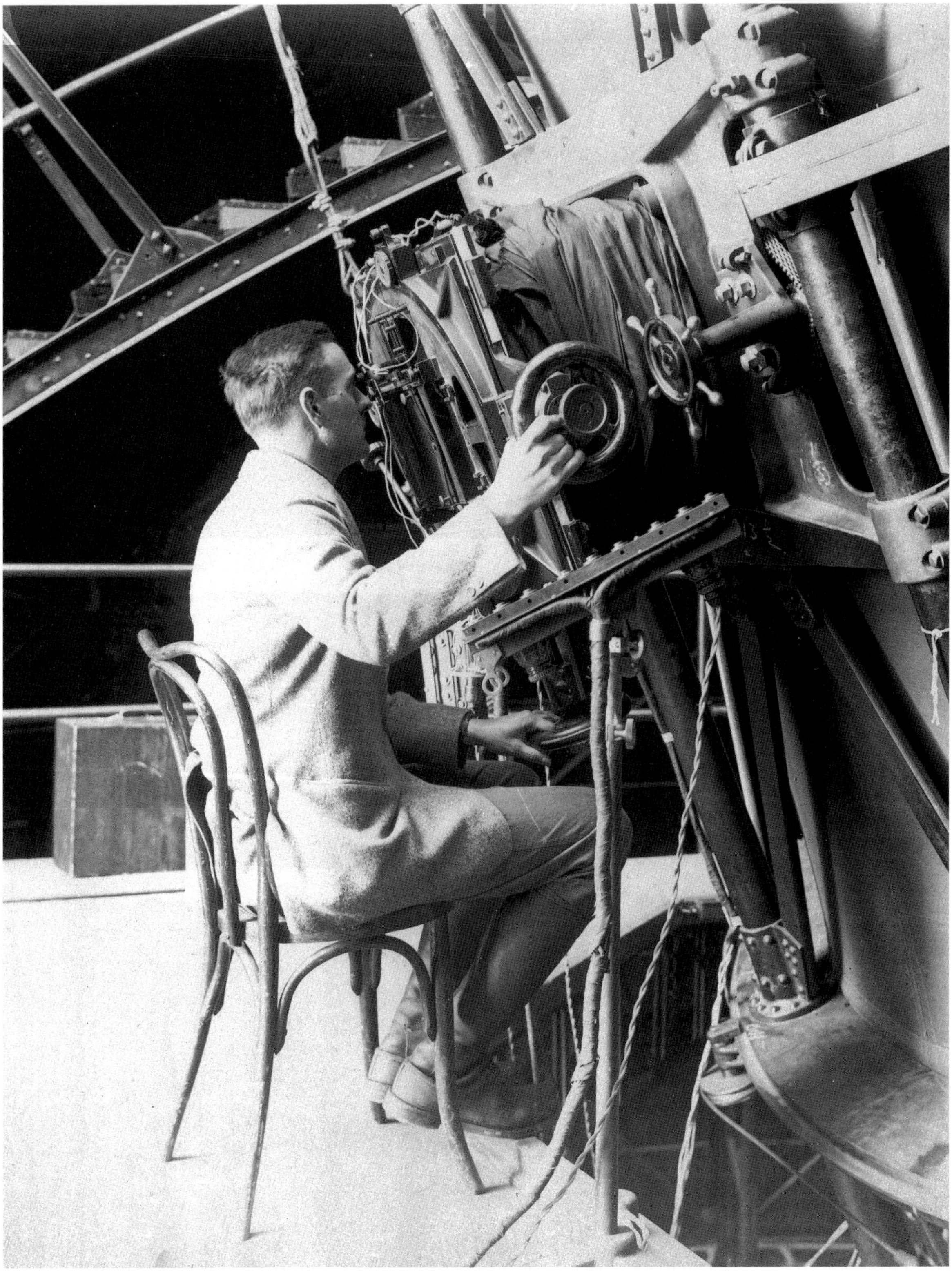

Edwin Hubble, ca. 1922, looking through the Newtonian focus of the 100-inch telescope at Mount Wilson. The combination of Hubble's extraordinary skills as an astronomer and the powers of the world's largest telescope was formidable, leading to some of the most important discoveries in the history of astronomy. (Huntington Library: Carnegie Institution of Washington Collection, COPC 2913)

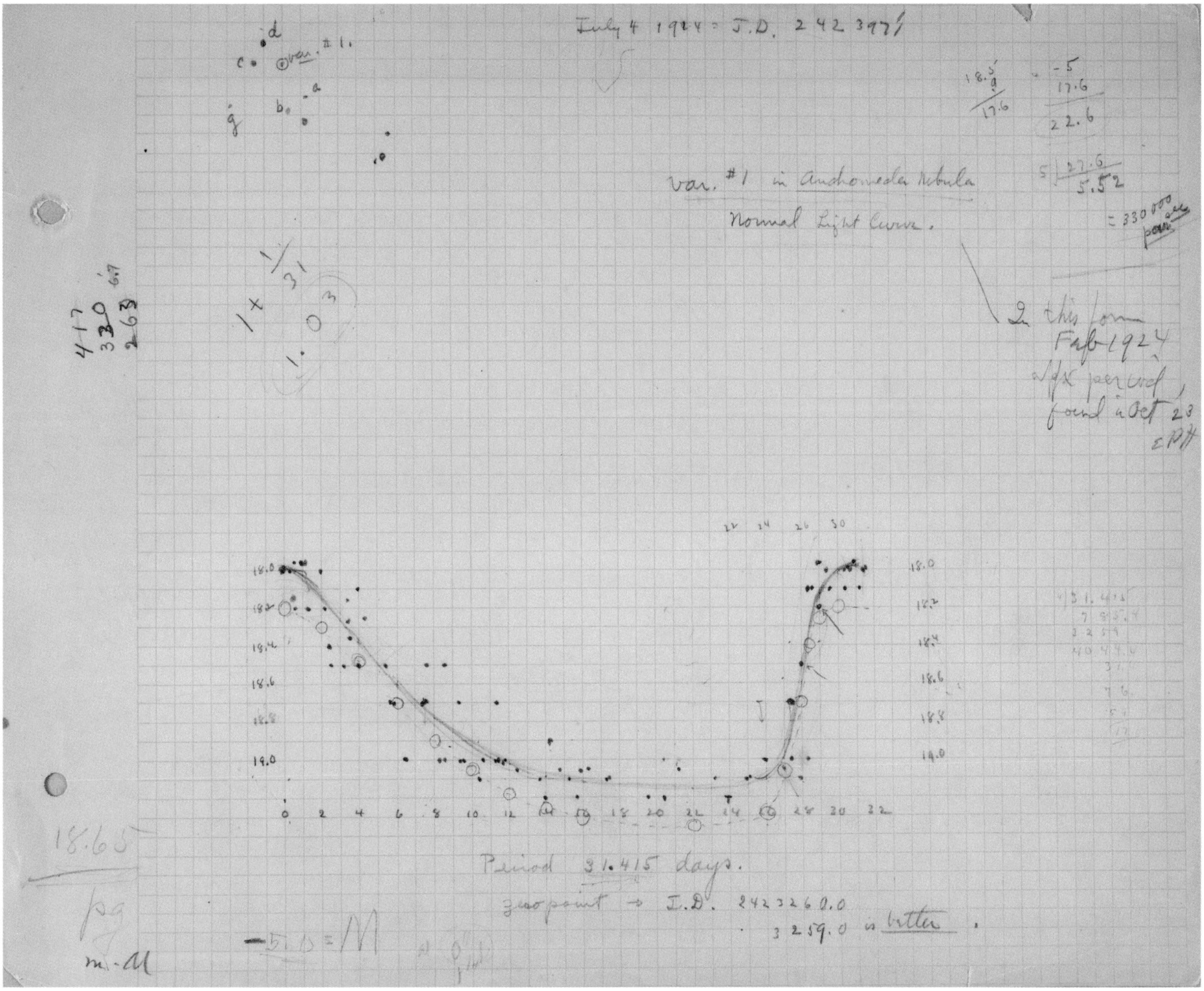

Edwin Hubble's handwritten chart of the light curve of the first Cepheid variable in the Andromeda Galaxy (M31), 1924. After observing the Cepheid variable star in the spiral nebula in Andromeda for a few months, Hubble was able to construct this graph which illustrates how the light of the star varied with a regular pattern. By timing how often the pattern repeated, Hubble was able to use the period-luminosity relationship to determine the distance to this spiral nebula and confirm that it did indeed exist beyond the limits of the Milky Way. (Huntington Library: Hubble Papers, Addenda, Box 1, Folder 2)

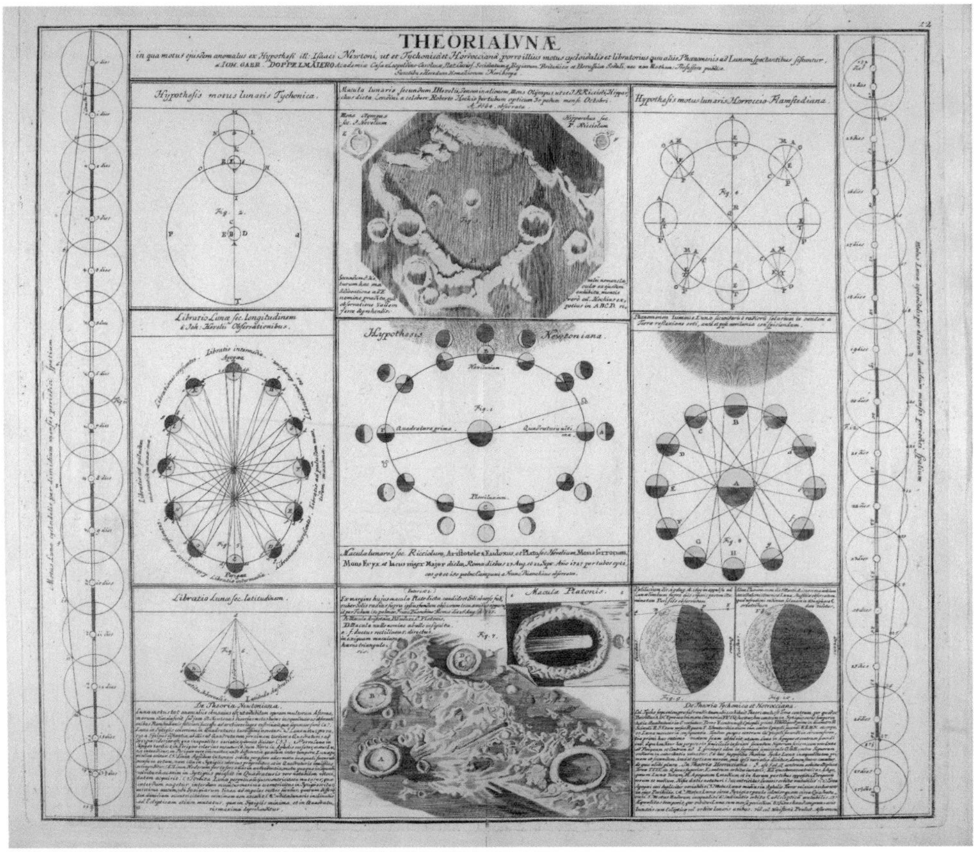
THEORIA LUNÆ
Hypothesis motus lunaris Tychonica.
Hypothesis motus lunaris Horroccio-Flamstediana.
Hypothesis Newtoniana.
Macula Platonis.

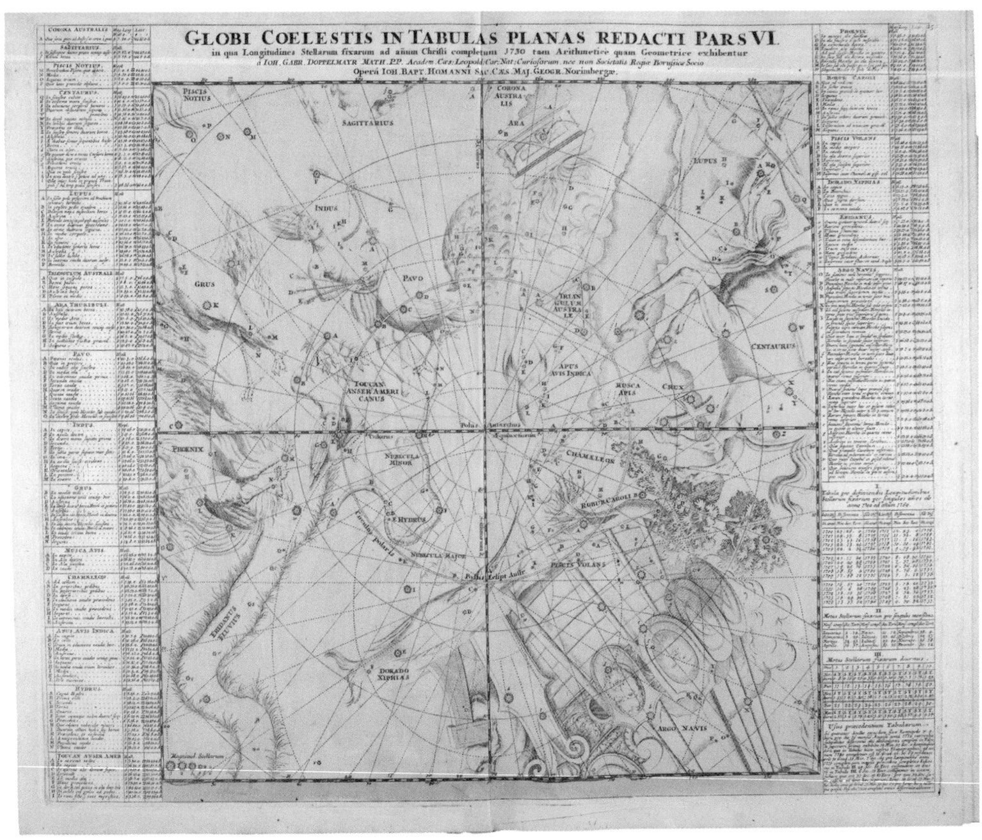
GLOBI COELESTIS IN TABULAS PLANAS REDACTI PARS VI.
SAGITTARIUS
INDUS
GRUS
PAVO
LUPUS
CENTAURUS
CRUX
PHŒNIX
CHAMÆLEON
PISCIS VOLANS
DORADO XIPHIAS
ARGO NAVIS

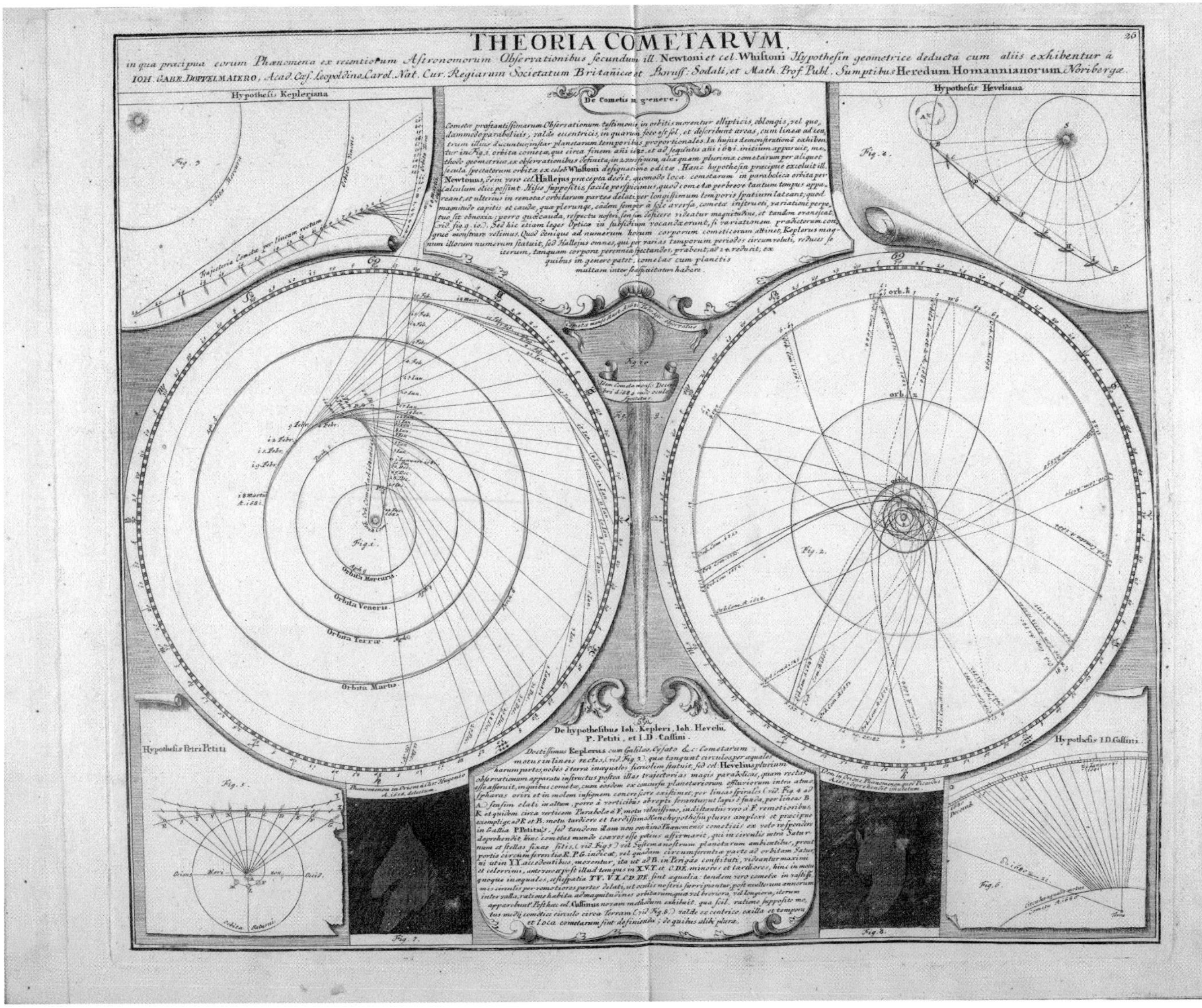

Gabriel Doppelmayr. *Atlas Coelestis* (Nuremberg: Heredum Homannianorum, 1742).

Opposite Top: Here Doppelmayr comparatively illustrates the ideas of fellow astronomers, demonstrating the theories of lunar motion put forth by Newton, Tycho, Flamsteed, and others.

Opposite Bottom: In this plate of his large atlas, Doppelmayr details the brightnesses of the stars as well as magnitude, longitude, and latitude of the stars within each constellation.

Above: This hand-colored illustration shows the observed paths of periodic comets and discusses the comet theories of such astronomers as Johannes Kepler, Johannes Hevelius, Petri Petiti, Jacques Cassini, and Edmund Halley. (Huntington Library: RB 487000.1084).

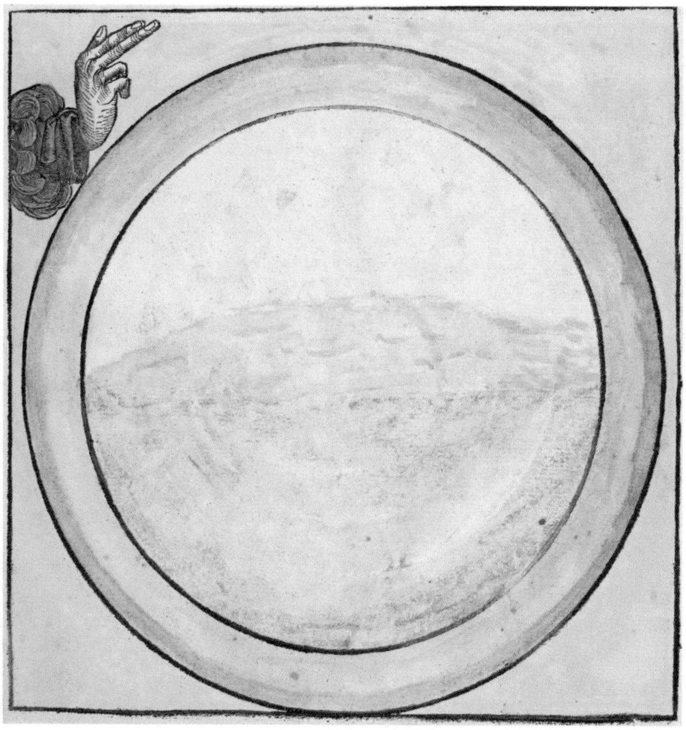

Hartmann Schedel. *Liber chronicarum* (Nuremberg: Antonius Koberger, 1493). Creation, days one through three. (Huntington Library: RB 95858)

Top left: Day one: The hand of God reaches forth from heaven to separate the void into two component parts of light and darkness.

Top right: Day two: Here God further separates the partially formed universe into the heavens (the outermost ring) and the waters below. No earth yet exists, and the illustrator has arbitrarily added extra rings to the illustration.

Bottom: On the third day of creation, dry land, water, and the heavens exist as the three rings of this illustration.

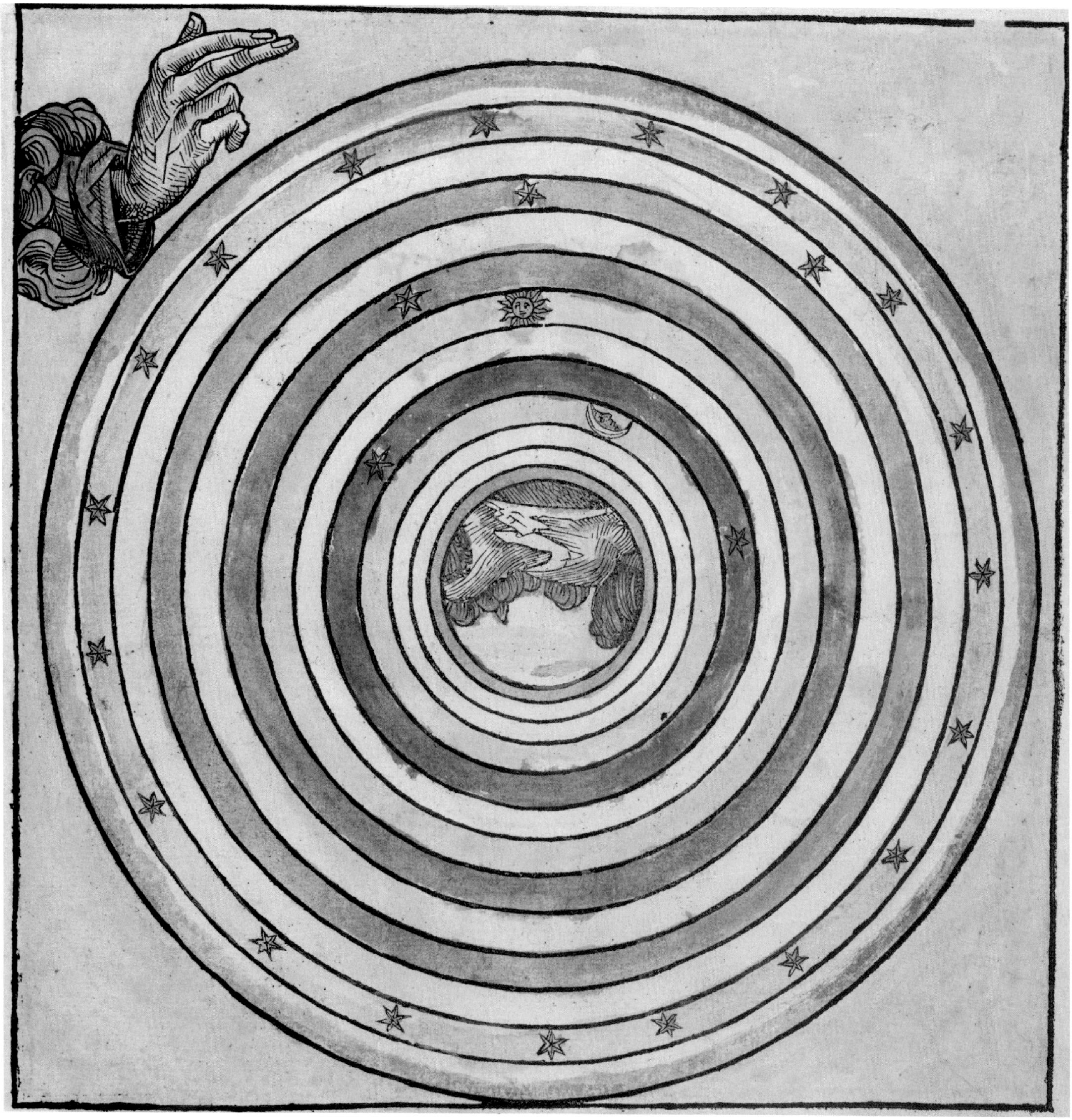

Hartmann Schedel. *Liber chronicarum* (Nuremberg: Antonius Koberger, 1493). On the fourth day, the celestial bodies begin to revolve around the Earth, at the center of the illustration, and the stars are recognizable as elements in the skies. The first rings circling Earth represent water, air, and fire. Then come seven spheres for the seven known planets—beginning with the Moon, and followed by Mercury, Venus, the Sun, Mars, Jupiter, and Saturn. The constellations comprise the next ring, followed by a series of fixed stars. Finally, the outer ring represents a concept of the era known as the mysterious *primum mobile* (Prime Mover), an engine of sorts that sets the finite apart from the infinite and keeps everything in uniform motion. (Huntington Library: RB 95858)

Hartmann Schedel. *Liber chronicarum* (Nuremberg: Antonius Koberger, 1493). The fifth day is represented by activities on Earth. It shows the creatures of land, sea, and sky that God made to populate the planet. (Huntington Library: RB 95858)

Hartmann Schedel. *Liber chronicarum* (Nuremberg: Antonius Koberger, 1493). On the sixth day, God, himself shows up on the scene and creates Adam from a lump of clay. (The seventh day of creation from the Nuremberg Chronicle is reproduced on page 48 of the volume.) (Huntington Library: RB 95858)

The Castle of Knowledge.
The Sphere of Destinye.
The wheele of Fortune.
Sphæra Fati
Sphæra Fortunæ.
QVI MODO SCANDIT, CORRVET STATIM.
whose gouernour is Knowledge.
whose ruler is Ignoraunce.
To KNOWLEDG is this Trophy set,
All learninges friendes will it support.
So shall their name great honour get,
And gaine great fame with good report.
Though spitefull Fortune turned her wheele
To staye the Sphere of Vranye,
Yet dooth this Sphere resist that wheele,
And fleeyth all fortunes villanye.
Though earthe do honour Fortunes balle,
And bytells blynde hyr wheele aduaunce,
The heauens to fortune are not thralle,
These Spheres surmount al fortunes chance.

RIGHT

Giambattista Riccioli. *Almagestum novum..., vol.* 2 (Bologna: Typographia haeredis Victorii Benatii, 1651). Riccioli's theory of the universe was a rival to the Tychonian conception. Riccioli disliked the fact that the orbit of Mars intersected that of the Sun in Tycho's conception of the universe. He thus rearranged the planets showing that Mars, Jupiter, and Saturn orbited the Earth, not the Sun. A "compromise theory" that differed from those of Ptolemy, Copernicus, and Tycho, it shows how astronomers struggled to fit the increasingly accurate data taken from celestial measurements with their theories of the universe. (Huntington Library: RB 487000.1030)

OPPOSITE

Robert Recorde. *The Castle of Knowledge* (London: Reginalde Wolfe, 1556). The title page of this English book is filled with symbolism. The figure on the left represents Destiny—well dressed and standing on a stable cube, holding a caliper and an armillary sphere, and studying the heavens, the repository for knowledge and truth. On the right stands Fate, standing on a wobbly ball, blindfolded, and holding the bridle of a horse and driving wheel—old technology. (Huntington Library: RB 69111)

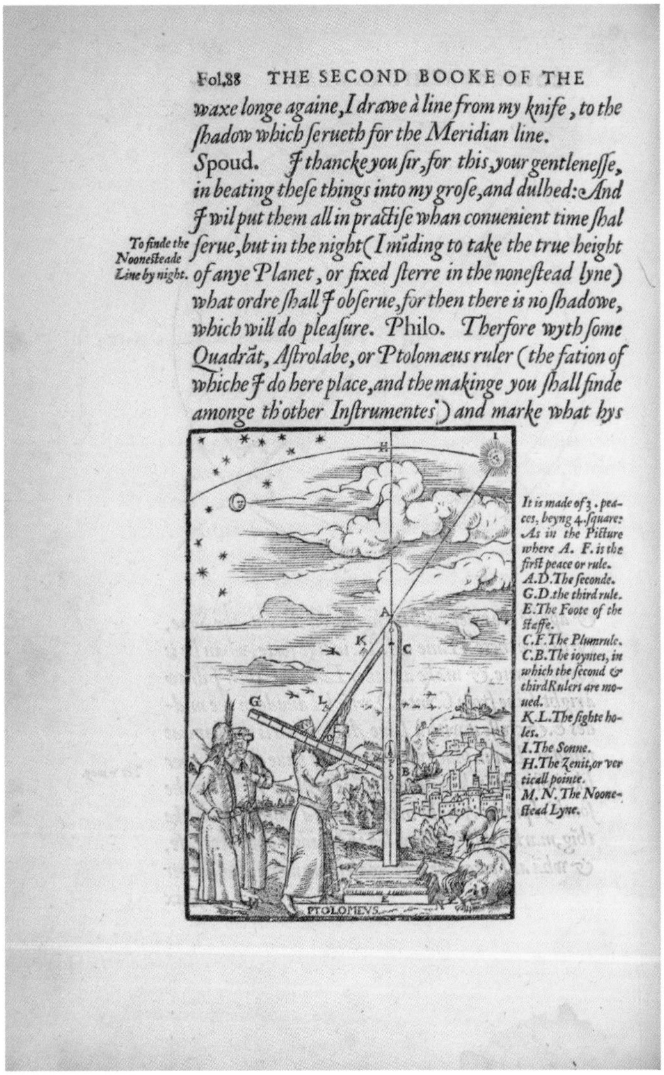
Fol.88 THE SECOND BOOKE OF THE

waxe longe againe, I drawe à line from my knife, to the shadow which serueth for the Meridian line.

Spoud. I thancke you sir, for this, your gentlenesse, in beating these things into my grose, and dulhed: And I wil put them all in practise whan conuenient time shal serue, but in the night (I mīding to take the true height of anye Planet, or fixed sterre in the nonestead lyne) what ordre shall I obserue, for then there is no shadowe, which will do pleasure. Philo. Therfore wyth some Quadrāt, Astrolabe, or Ptolomæus ruler (the fation of whiche I do here place, and the makinge you shall finde amonge th'other Instrumentes) and marke what hys

To finde the Noonesteade Line by night.

It is made of 3. peaces, beyng 4. square: As in the Picture where A. F. is the first peace or rule. A.D. The seconde. G.D. the third rule. E. The Foote of the staffe. C.F. The Plumrule. C.B. The ioyntes, in which the second & third Rulers are moued. K.L. The sighte holes. I. The Sonne. H. The Zenit, or verticall pointe. M. N. The Noonestead Lyne.

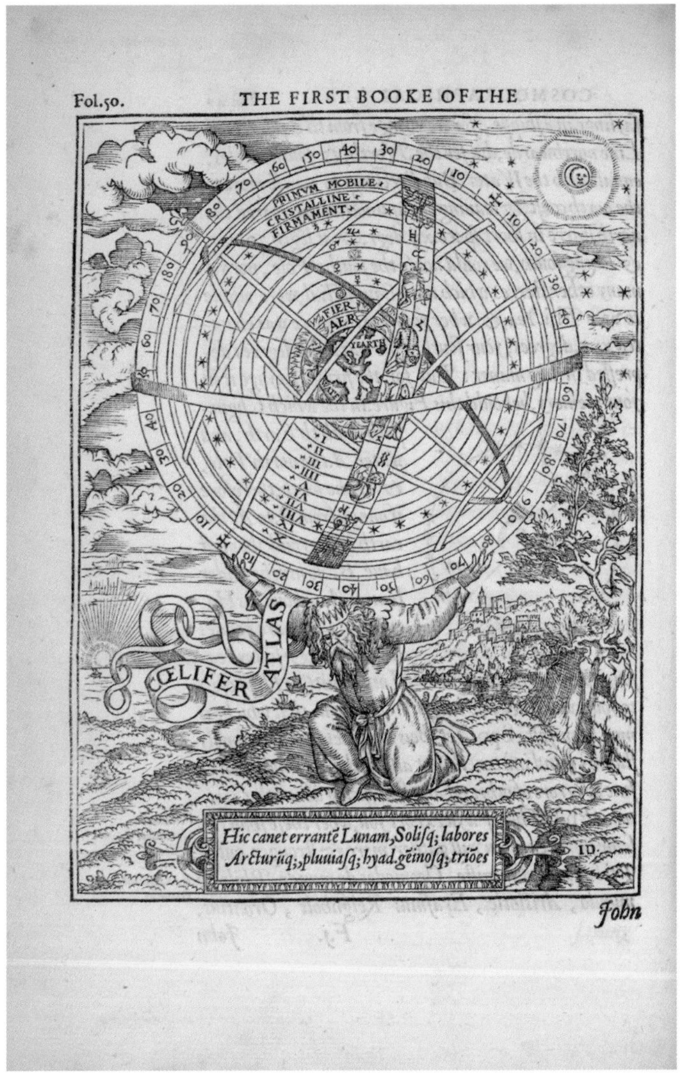

William Cuningham. *The cosmographical glasse: conteinyng the pleasant principles of cosmographie, geographie, hydrographie, or navigation* (London: In officina Ioan. Day typographi, 1559).

Left: This illustration shows the use of a measuring staff, designed to take the angle between a point directly overhead and the Sun. Measurement devices before the telescope included the cross-staff, Jacob's staff, the sextant, the astrolabe, and other creations that allowed for increasingly precise measurements of the degrees between—and thus the position of—the stars and planets relative to various fixed points.

Right: Atlas bearing the heavens on his shoulders in the form of an armillary sphere. Surrounding Earth and outer rings of air and fire are the seven planetary spheres, the firmament of fixed stars, and the crystalline sphere. (Huntington Library: RB 60873).

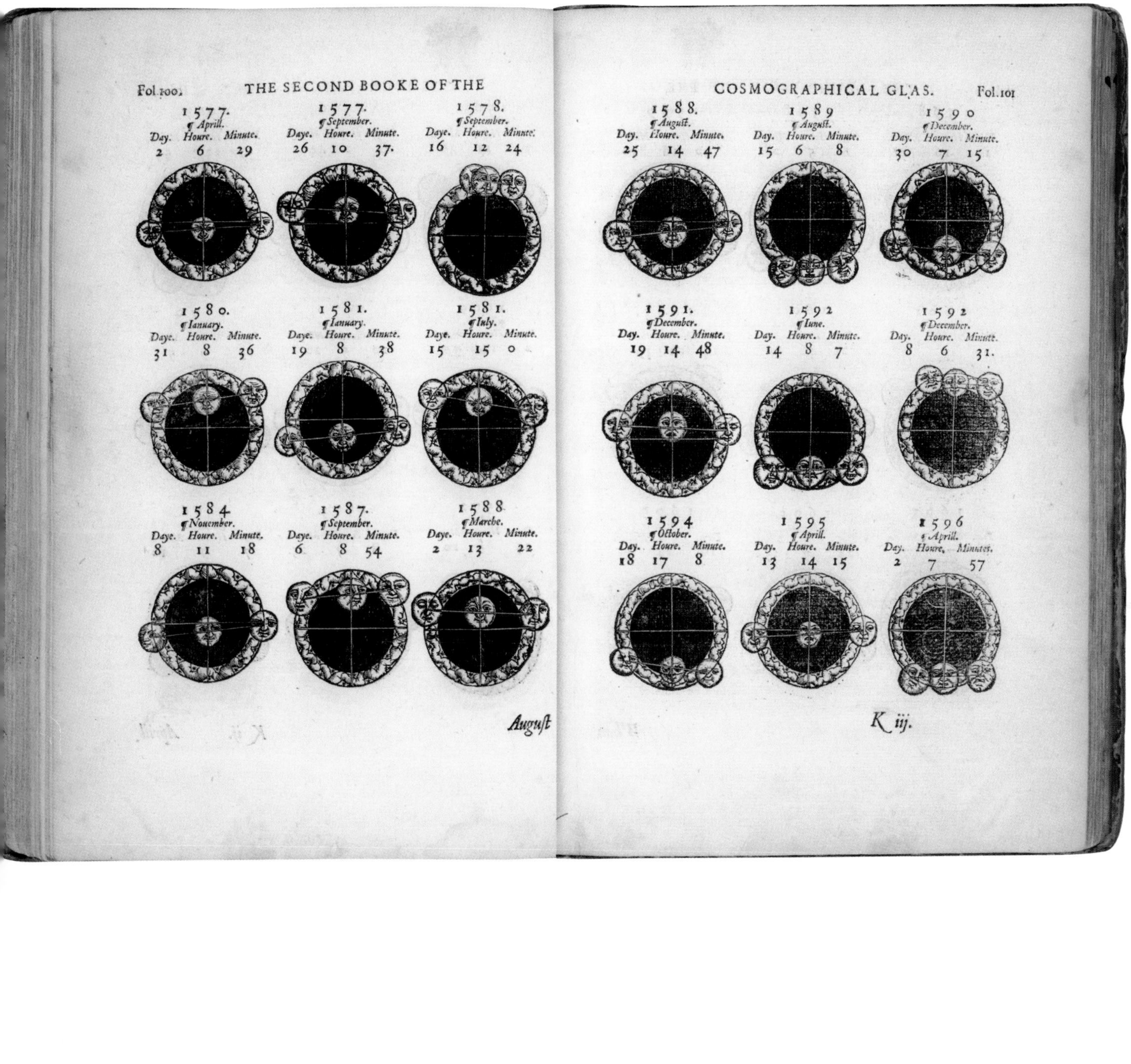

William Cuningham. *The cosmographical glasse: conteinyng the pleasant principles of cosmographie, geographie, hydrographie, or navigation* (London: In officina Ioan. Day typographi 1559). Cuningham calculated the date of lunar eclipses taking place between March 1560 and September 1605. In some years the Moon is shown smiling; in others, frowning or sleepy. (Huntington Library: RB 60873).

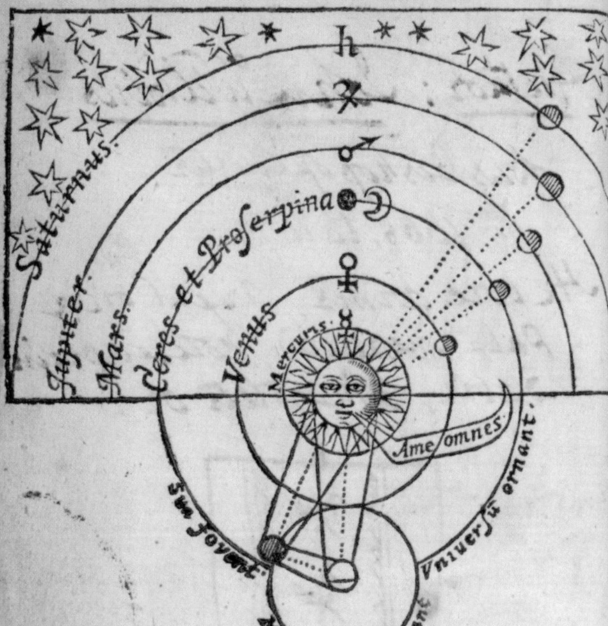

THE DISCOVERY OF A WORLD IN THE MOONE.

OR,

A DISCOVRSE Tending TO PROVE, that 'tis probable there may be another habitable World in that Planet.

Quid tibi inquis ista proderunt? Si nihil aliud, hoc certè, sciam omnia angusta esse. SENECA. Præf. ad 1. Lib. N. Q.

By Bp Wilkins.

LONDON,
Printed by E.G. for *Michael Sparke* and *Edward Forrest*. 1638.

FIVE LIFE ON OTHER WORLDS

John Wilkins. *Discovery of a World in the Moone* (1st edition. London: Printed by E.G. for Michael Sparke..., 1638). Although sometimes portrayed as an eccentric seventeenth-century Jules Verne, the Oxford scholar John Wilkins was open-minded, enthusiastic, and successful in winning favor in high places. This not only made him a top academic politician, but also an important agent in disseminating and institutionalizing the new science of the day. (Huntington Library: RB 600935).

BERNARD FONTENELLE, a best-selling seventeenth-century writer on the possibilities of life beyond Earth, noted wryly that "All philosophy is based on two things only: curiosity and poor eyesight." It is true that early telescopes were not (and, of course, still are not) powerful enough to resolve decisively such issues as the existence of life on other planets. But that has never prevented speculation about the possibility. The religious implications of such theories limited the number that were proposed in early times. Articulating such philosophies could be hazardous to one's health.[2]

The first two printed works on the possibility of travel to other worlds and life on those planets were both published in England in 1638. The first of these, although given less credit as an early work on extraterrestrial life and travel, had a number of progressive ideas. Francis Godwin of Hereford (1552–1633), who, as a student at Oxford in 1583, had been influenced by ideas propounded by Giordano Bruno (1548–1600), wrote a literary work of fantasy entitled *The Man in the Moone, or, a Discourse of a Voyage Thither.* (Bruno's works included a book entitled *On the Infinite Universe and Worlds,* which proposed that all the planets of the solar system were, like the Earth, inhabited.) As the title of Godwin's work implies, his book discusses possible means of actually getting to the Moon as well as the extraterrestrial life that might be found there. This idea of travel to other worlds and the possibilities of life there was not entirely new: Galileo, Descartes, and Kepler had all cautiously recognized the possibility of extraterrestrial life. But Godwin raised the two issues of travel and life decisively. His ideas were extraordinarily progressive for a seventeenth-century literary figure who was not technically or scientifically trained. Godwin's romance is filled with conceptions of gravity and weightlessness, and hints of the helicopter, nuclear power, reverse thrust, and non-human payloads, ideas that have evolved in the modern phase of aeronautics. It remained steadily in print from its first publication in 1638 (only one copy of this first edition survives, at the British Museum) through 1768, appearing in at least twenty-five editions. The work influenced writers of both literature and science: Cyrano de Bergerac (1619–1655) based his "A Voyage to the Moon" on Godwin's work (as well as on that of John Wilkins); Bernard Fontenelle, Christiaan Huygens, Daniel Defoe, Jonathan Swift, and many other writers and astronomers borrowed richly from it.

The English religious figure and natural philosopher John Wilkins (1614–1672), simultaneously proposed manned flight to the Moon in his 1638 book *Discovery of a World in the Moone: Or A Discourse Tending To Prove that 'tis probable there may be another habitable World in that Planet.* A Protestant clergyman who later became a bishop in the Anglican church, Wilkins thought it "probable" that a habitable world existed on the Moon and was eager to show that such a concept did not contradict either faith or reason. Belief in a lunar world, he insisted, advanced the divine wisdom by "shewing a compendium of providence, that could make the same body a world, and a Moone; a world for habitation, and a Moone for the use of others, and the ornament of the whole frame of Nature." He theorized that the dark spots on the

Moon were seas and the lighter regions were solid ground and discussed the possibility of an atmosphere on the Moon. His work was bound by a consistent but incorrect logic: "If our Earth were one of the Planets [according to the Copernicans] why may not another of the Planets be an Earth?"

Wilkins's work was very popular, as seven printings between 1638 and 1802 attest. It showed tremendous learning for a man so young (he wrote the first edition when he was only twenty-four), integrating the views of a wide variety of scholars to make his various points. "'Tis not Aristotle, but truth that should be the rule of our opinions," he remarked in his attempts to undermine the foundations of Aristotelian physics. The work was translated into French in 1655 as *Le monde dans la lune*, and published in Rouen, the hometown of Bernard Fontenelle, who played an important role in popularizing the idea that life existed on other worlds.

Bernard Fontenelle, who lived to be a month shy of a century old (1657–1757) and was not yet born when Wilkins's work was translated into French, likely had access to Wilkins's book and was probably influenced by it to some degree. Talented, polished, and charming, Fontenelle first penned a comedy about comets, *La comète* (1681), in which he tried to ease widespread worries about comets stirred up by the fiery comet of 1680, which had caused considerable panic and was considered a bad omen. Fontenelle wrote *Entretiens sur la pluralité des mondes* in Paris in 1686 (translated into English the following year as *A Discourse on the Plurality of Worlds*). Strongly influenced by Copernicus, Galileo, Francis Bacon, and Descartes, he was eager to promote a heliocentric version of the universe. He wrote with verve, charm, and flippancy, yet with a carefully laid-out dialogue between his narrator and a female student. He promoted extremely unorthodox views in the work; he was never heavy-handed or dogmatic, and although his book ended up on the Catholic Church's list of banned books, he was never persecuted. Fontenelle, a man of letters, served perennially as the secretary of the French Académie des Sciences and reached great popular success writing on the subject; approximately one hundred different editions of his work have been published since *Entretiens* first appeared.

Another man who advanced the idea of extraterrestrial intelligence in the seventeenth century was the great Dutch physicist and astronomer Christiaan Huygens (1629–1695), one of the leading thinkers of his day. Huygens made important advances in the fields of physics, mathematics, optics, and astronomy. He developed the idea of making practical use of a pendulum as a means of regulating clockworks, advanced the study of light waves and gravity, and discovered the Saturnian moon Titan. In 1698 a work of his known as the *Cosmotheoros* was published posthumously under the title *The Celestial Worlds Discover'd: or, Conjectures Concerning the Inhabitants, Plants and Productions of the Worlds in the Planets.* In it, he sets out a methodical argument—because the Copernican view of the universe revealed that Earth held no privileged position in the heavens, it was unreasonable to suppose that life should be restricted to Earth alone. The providence of God and his infinite wisdom played a fascinating central role in Huygens's reasoning. He postulated that here must be life on other planets and living entities capable of reason because, in their absence, this creation would be senseless and the Earth, once again, would have an unreasonably privileged position. Writing from the more exalted pulpit of the scientist, Huygens gave added credibility to the idea of life on other worlds.

By 1750 the Copernican revolution had established that our planets revolved around a sun, and made a strong case for the idea that other stars existed and that these were probably suns with their own planets—a change in thinking that had profound implications. If life exists in our solar system, might it not reasonably exist

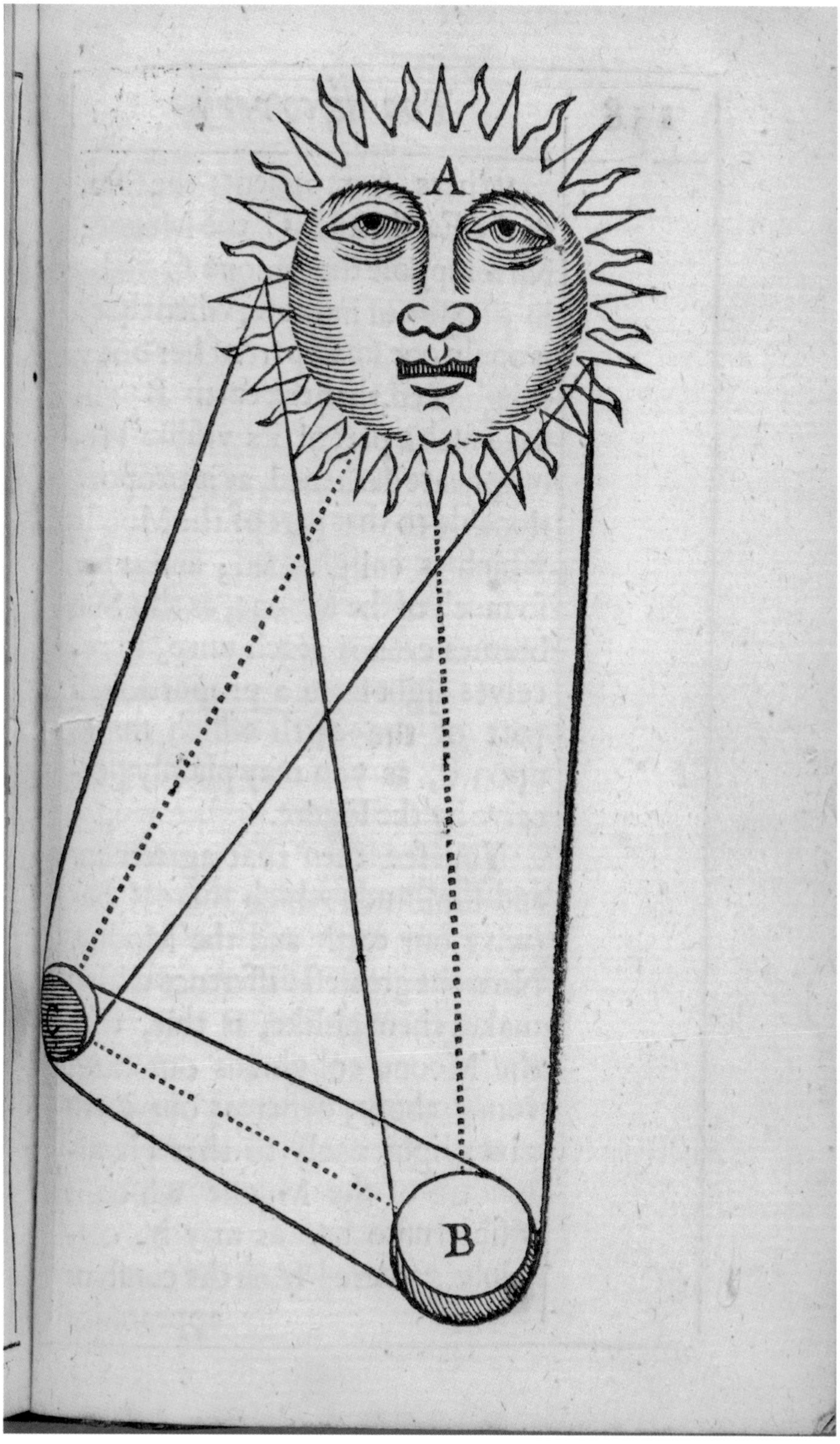

John Wilkins. *Discovery of a World in the Moone* (1st edition. London: Printed by E.G. for Michael Sparke, 1638). This illustration shows how the Sun's light reflects from the Moon and is visible on earth. The Sun was often personified in this manner, but the same treatment was very rarely applied to the Earth. (Huntington Library: RB 600935).

in what appeared to be countless other solar systems? A number of other very important writers added support to the idea of a plurality of worlds in the eighteenth century, including Thomas Wright, one of the pioneers of stellar astronomy; William Herschel, the famous British astronomer who discovered Uranus; and a number of others, including John Locke, Voltaire, Gottfried Leibniz, and Immanuel Kant. The situation began to change somewhat by the mid-nineteenth century, however, in a backlash against the overconfident pluralism of the period. The most important of those who argued against this was by William Whewell, who published an essay in 1853 entitled *Of the Plurality of Worlds,* a work that argued that the conditions probable on other planets in our solar system would almost certainly preclude life comparable to humans. But hope and imagination both ran high. Writer Percival Lowell (1855–1916) suggested that its canals and polar icecaps made Mars the best candidate for life elsewhere in the solar system, earning him the role of patron saint for the "life-on-Mars" movement popular in the nineteenth century. It was a hopeful dream of many stargazers, but one lacking proof. Publisher William Randolph Hearst is said to have queried a famous astronomer of his day with the message: "Is there life on Mars? Please cable one thousand words." The astronomer's response was "Nobody knows"—repeated five hundred times.

A curious duality had arisen by the start of the twentieth century: reason dictated more and more strongly that life must exist on other worlds, but not a scrap of proof existed. This allowed for a rich imaginative literature to arise; and science fiction became a genre of increasing popularity. In 1901 Herbert George (H. G.) Wells wrote *The First Men in the Moon,* a cautionary tale about visiting other worlds. Wells had a fairly firm grounding in science, having attended the Royal College of Science where he studied with the biologist Thomas Henry Huxley. His novel on life on the Moon marked a shift from works based on science and evolutionary ideas to notions of politics and sociology. Much of *The First Men in the Moon* satires the imperialism of the time, using a visit to the Moon as representative of human ideas of conquest and foreign penetration. *DL*

OPPOSITE

John Wilkins. *Discovery of a World in the Moone* (1st edition. London: Printed by E.G. for Michael Sparke, 1638). This diagram is part of an argument constructed by Wilkins to show, by using Earth as an example, that the Moon—and other worlds—might contain their own sphere of meteors and other astronomic activity. He notes by explanation that "When there happens to be a concourse of some few starres, then doe many other flie unto them from all the parts of heaven like so many Bees unto their King." (Huntington Library: RB 600935).

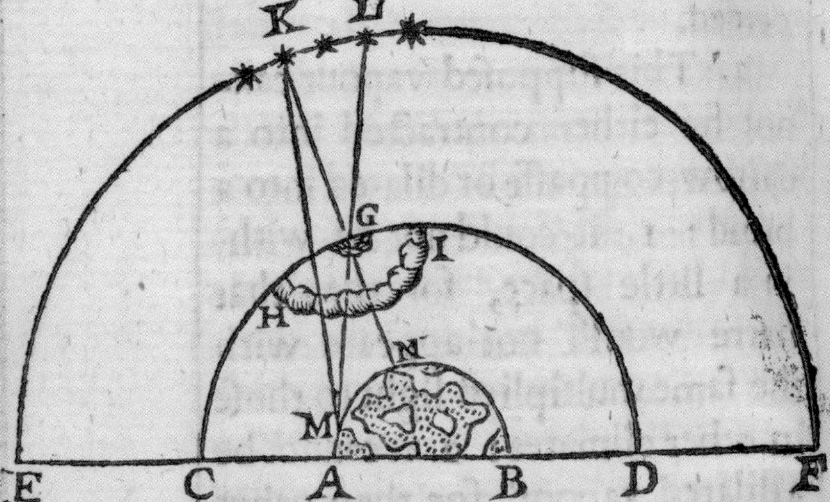

Suppose A B to be a Hemi-spheare of one earth, C D to be the upper part of the highest region, in which there might be either a contracted vapour, as G, or else a dilated one, as H I. Suppose E F likewise to represent halfe the heavens, wherein was this appearing Comet at K. Now I say, that a contracted vapour, as G could not cause this appearance, because an inhabitant at M could not discerne the same starre with

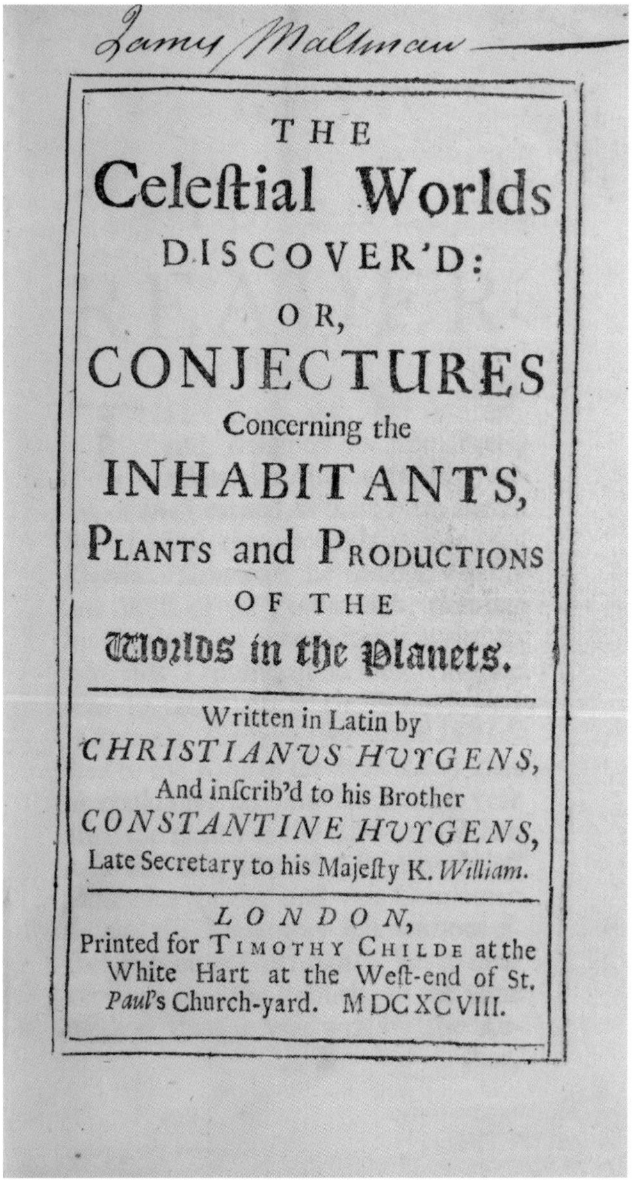

THE
Celeftial Worlds
DISCOVER'D:
OR,
CONJECTURES
Concerning the
INHABITANTS,
PLANTS and PRODUCTIONS
OF THE
Worlds in the Planets.

Written in Latin by
CHRISTIANUS HUYGENS,
And infcrib'd to his Brother
CONSTANTINE HUYGENS,
Late Secretary to his Majefty K. *William.*

LONDON,
Printed for TIMOTHY CHILDE at the
White Hart at the Weft-end of St.
Paul's Church-yard. MDCXCVIII.

Christiaan Huygens. *The Celestial Worlds Discover'd: or, Conjectures Concerning the Inhabitants, Plants and Productions of the Worlds in the Planets* (London: Printed for Timothy Childe..., 1698). Huygens not only advanced methods for making telescope lenses more powerful but also identified Saturn's rings and one of Saturn's moons, promoted the theory of light as a wave, and undertook important work on pendulum clocks. He is less well known for his work on life on other planets, but it was finally this work that publicly validated the idea of extraterrestrial life. (Huntington Library: RB 235132)

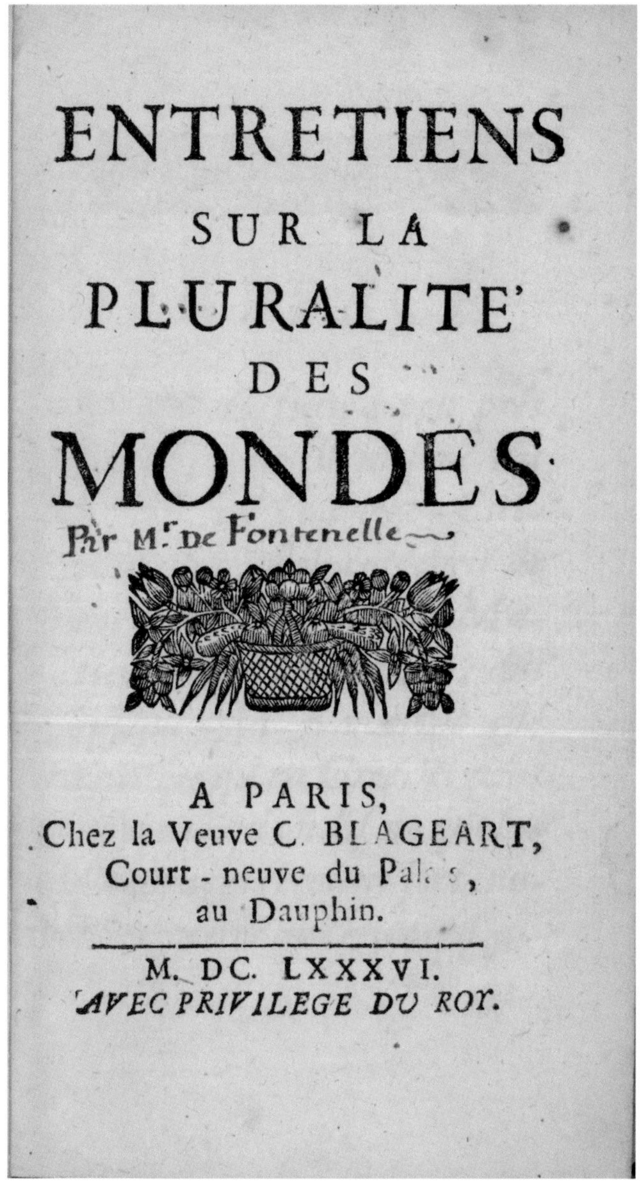

ENTRETIENS
SUR LA
PLURALITE'
DES
MONDES
Par M.r De Fontenelle

A PARIS,
Chez la Veuve C. BLAGEART,
Court - neuve du Palais,
au Dauphin.

M. DC. LXXXVI.
AVEC PRIVILEGE DU ROY.

Bernard Fontenelle. *Entretiens sur la pluralité des mondes* (1st edition. Paris: Chez la Veuve C. Blageart, 1686). Fontenelle, a popular French writer, pushed into dangerous territory with this somewhat heretical work; it ended up on the Catholic church's list of banned books. "If I had my hand full of truth, I would take good care how I opened it," the witty and articulate Fontenelle noted wryly. (Huntington Library: RB 381966).

"Insects," murmured Cavor, "insects"

L'HOMME
DANS
LA·LUNE

Herbert George Wells. *The First Men in the Moon* (1st edition. London: George Newnes, Ltd., 1901). A prophetic description of the methodology of space flight, the novel contained a thinly veiled satire of the growing U.S. imperialism of the time. Although Wells's novels were highly entertaining, he also tried to pave the way for a wiser attitude about the future of the mankind. (Huntington Library: RB 181753)

Francis Godwin. *Le Homme dans la lune...* (Paris: Jean Cochart, 1671). In this French edition of Godwin's work, which appeared in English in 1638, travel to the Moon is projected. The tale records the adventures of a Spanish sailor named Domingo Gonsales, who trained swans for an eventful trip that was to take him eleven days. (Huntington Library: RB 4073)

CANON MOTVVM LONGITVDINIS VENERIS, Apogæi, & Nodi eiusdem in annis expansis.

Anni	Medius motus Sig.	Gr.	′	″	Apogæi Gr.	′	″	Nodi Gr.	′	″
21	1	18	40	19	0	27	19	0	16	27
22	9	3	27	55	0	28	37	0	17	14
23	4	18	15	31	0	29	55	0	18	1
B 24	0	4	39	15	0	31	13	0	18	48
25	7	19	26	52	0	32	31	0	19	35
26	3	4	14	28	0	33	49	0	20	22
27	10	19	2	4	0	35	7	0	21	9
B 28	6	5	25	48	0	36	25	0	21	56
29	1	20	13	24	0	37	43	0	22	43
30	9	5	1	0	0	39	1	0	23	30
31	4	19	48	37	0	40	19	0	24	17
B 32	0	6	12	20	0	41	37	0	25	4
33	7	20	59	57	0	42	55	0	25	51
34	3	5	47	33	0	44	13	0	26	38
35	10	20	35	9	0	45	31	0	27	25
B 36	6	6	58	53	0	46	49	0	28	12
37	1	21	46	29	0	48	7	0	29	9
38	9	6	34	5	0	49	25	0	29	46
39	4	21	21	42	0	50	43	0	30	33
B 40	0	7	45	26	0	52	1	0	31	20
41	7	22	33	2	0	53	20	0	32	7
42	3	7	20	38	0	54	38	0	32	54
43	10	22	58	14	0	55	56	0	33	41
B 44	6	8	31	58	0	57	14	0	34	28
45	1	23	19	34	0	58	32	0	35	15
46	9	8	7	10	0	59	50	0	36	2
47	4	22	54	47	1	1	8	0	36	49
B 48	0	9	18	31	1	2	26	0	37	36
49	7	24	6	7	1	3	44	0	38	23
50	3	8	53	43	1	5	2	0	39	10
51	10	23	41	20	1	6	20	0	39	57
B 52	6	10	5	3	1	7	38	0	40	44
53	1	24	52	39	1	8	56	0	41	31
54	9	9	40	15	1	10	14	0	42	18
55	4	24	27	52	1	11	32	0	43	5
B 56	0	10	51	36	1	12	50	0	43	52
57	7	25	39	12	1	14	8	0	44	39
58	3	10	26	48	1	15	26	0	45	26
59	10	25	14	24	1	16	44	0	46	13
B 60	6	11	38	8	1	18	3	0	47	0
61	1	26	25	45	1	19	21	0	47	47
62	9	11	13	21	1	20	39	0	48	34
63	4	26	0	57	1	21	57	0	49	21
B 64	0	12	24	41	1	23	15	0	50	8
65	7	27	12	17	1	24	33	0	50	55
66	3	11	59	53	1	25	51	0	51	42
67	10	26	47	30	1	27	9	0	52	29
B 68	6	13	11	14	1	28	27	0	53	16
69	1	27	58	50	1	29	45	0	54	3
70	9	12	46	26	1	31	3	0	54	50
71	4	27	34	2	1	32	21	0	55	37
B 72	0	13	57	46	1	33	39	0	56	24
73	7	28	45	22	1	34	57	0	57	11
74	3	13	32	58	1	36	15	0	57	58
75	10	28	20	35	1	37	33	0	58	45
B 76	6	14	44	19	1	38	51	0	59	32
77	1	29	31	55	1	40	9	1	0	19
78	9	14	19	31	1	41	27	1	1	6
79	4	29	7	7	1	42	45	1	1	53
B 80	0	15	30	51	1	44	4	1	2	40
81	8	0	18	27	1	45	22	1	3	27
82	3	15	6	3	1	46	40	1	4	14
83	10	29	53	40	1	47	58	1	5	1
B 84	6	16	17	24	1	49	16	1	5	48
85	2	1	5	0	1	50	34	1	6	35
86	9	15	52	36	1	51	52	1	7	22
87	5	0	40	12	1	53	10	1	8	9
B 88	0	17	3	56	1	54	28	1	8	56
89	8	1	51	32	1	55	46	1	9	43
90	3	16	39	8	1	57	4	1	10	30
91	11	1	26	45	1	58	22	1	11	17
B 92	6	17	50	29	1	59	40	1	12	4
93	2	2	38	5	2	0	58	1	12	51
94	9	17	25	41	2	2	16	1	13	38
95	5	2	13	17	2	3	34	1	14	25
B 96	0	18	37	1	2	4	52	1	15	12
97	8	3	24	38	2	6	10	1	15	59
98	3	18	12	14	2	7	28	1	16	46
99	11	2	59	50	2	8	46	1	17	33
100	6	19	23	34	2	10	5	1	18	20

CANON MOTVVM LONGITVDINIS VENERIS, APOGÆI AC NODI eiusdem in mensibus anni Communis, & Bissextilis.

Anni Commun.	Medius motus Sig.	Gr.	′	″	Apoggi ′	″	Nodi ′	″
Ianuarius	1	19	40	3	0	6	0	4
Februar.	3	4	31	42	0	12	0	7
Martius	4	24	11	44	0	19	0	11
Aprilis	6	12	15	39	0	25	0	15
Maius	8	1	55	42	0	31	0	19
Iunius	9	19	57	37	0	38	0	23
Iulius	11	9	39	39	0	45	0	27
Augustus	0	29	19	42	0	52	0	31
Septemb.	2	17	23	37	0	58	0	35
October	4	7	3	39	1	5	0	39
Nouemb.	5	21	7	34	1	11	0	43
Decemb.	7	14	47	36	1	18	0	47

Anni Bissextilis	Medius motus Sig.	Gr.	′	″	Apoggi ′	″	Nodi ′	″
Ianuarius	1	19	40	3	0	6	0	4
Februar.	3	6	7	50	0	12	0	7
Martius	4	25	47	52	0	19	0	11
Aprilis	6	13	51	47	0	25	0	15
Maius	8	3	31	50	0	32	0	19
Iunius	9	21	33	45	0	38	0	23
Iulius	11	11	35	47	0	45	0	27
Augustus.	1	0	55	50	0	52	0	31
Septemb	2	18	59	45	0	58	0	35
October	4	8	39	47	1	5	0	39
Nouemb.	5	22	41	42	1	11	0	43
Decemb.	7	16	23	44	1	18	0	47

Canon motuum longitudinis Veneris Apogæi ac Nodi eiusdem in diebus.

	Medius motus Sig.	Gr.	′	″	Apoggi ′	″	Nodi ′	″
1	0	1	36	8	0	0	0	0
2	0	3	12	16	0	0	0	0
3	0	4	48	23	0	1	0	0
4	0	6	24	31	0	1	0	0
5	0	8	0	39	0	1	0	1
6	0	9	36	47	0	1	0	1
7	0	11	12	55	0	1	0	1
8	0	12	49	3	0	2	0	1
9	0	14	25	10	0	2	0	1
10	0	16	1	18	0	2	0	1
11	0	17	37	26	0	2	0	1
12	0	19	13	34	0	3	0	2
13	0	20	49	42	0	3	0	2
14	0	22	25	50	0	3	0	2
15	0	24	1	57	0	3	0	2
16	0	25	38	5	0	3	0	2
17	0	27	14	13	0	4	0	2
18	0	28	50	21	0	4	0	2
19	1	0	26	29	0	4	0	2
20	1	2	2	37	0	4	0	3
21	1	3	38	44	0	4	0	3
22	1	5	14	52	0	5	0	3
23	1	6	51	0	0	5	0	3
24	1	8	27	8	0	5	0	3
25	1	10	3	16	0	5	0	3
26	1	11	39	23	0	6	0	3
27	1	13	15	31	0	6	0	3
28	1	14	51	39	0	6	0	4
29	1	16	27	47	0	6	0	4
30	1	18	3	55	0	6	0	4
31	1	19	40	3	0	7	0	4

Canon motus longitudinis Veneris in horis, & horarum scrupulis.

H. ′ ″	Gr. ′ ″	′ ″ ‴	″ ‴ ⁗	H. ′ ″	Gr. ′ ″	′ ″ ‴	″ ‴ ⁗
1	0	4	0	31	2	4	12
2	0	8	1	32	2	8	12
3	0	12	1	33	2	12	12
4	0	16	2	34	2	16	13
5	0	20	2	35	2	20	13
6	0	24	2	36	2	24	13
7	0	28	3	37	2	28	14
8	0	32	3	38	2	32	14
9	0	36	3	39	2	36	14
10	0	40	4	40	2	40	15
11	0	44	4	41	2	44	15
12	0	48	5	42	2	48	15
13	0	52	5	43	2	52	16
14	0	56	5	44	2	56	16
15	1	0	6	45	3	0	16
16	1	4	6	46	3	4	17
17	1	8	6	47	3	8	17
18	1	12	7	48	3	12	17
19	1	16	7	49	3	16	18
20	1	20	8	50	3	20	18
21	1	24	8	51	3	24	18
22	1	28	8	52	3	28	19
23	1	32	9	53	3	32	19
24	1	36	9	54	3	36	20
25	1	40	9	55	3	40	20
26	1	44	10	56	3	44	20
27	2	48	10	57	3	48	21
28	1	52	11	58	3	52	21
29	1	56	11	59	3	56	21
30	2	0	11	60	4	0	22

SIX | TABLES AND CELESTIAL MECHANICS

Vincenzo Renieri, *Tabulae motuum caelestium universales Serenissimi Magni Ducis Etruriae Ferdinandi II*... (Florence: Amatore Massa, 1647). Second edition. A page illustrating a portion of Kepler's Rudolphine Tables of planetary motion. (Huntington Library: RB 476466)

ONE OF THE PRIMARY GOALS of early astronomers was the determination of the exact locations of the Sun, Moon, and planets in the sky. Being able to know the current positions and predict the future movements would enable astronomers to determine such things as when eclipses would occur as well as cast horoscopes for astrological purposes. Ptolemy seems to have originated the development of unified planetary tables that would allow him to predict the location of any planet on any given date. Using the mathematical model of the universe elucidated in his *Almagest*, Ptolemy constructed his "Handy Tables" of planetary positions around 150 C.E. The Handy Tables became the foundation for the tables produced by the Islamic astronomers. These Islamic astronomical tables (or *zījes*) became quite common, and a number of noted astronomers produced them. They became quite well known to the Western Europeans, with the *zījes* of Al-Battānī actually being published in 1537. In Toledo, Spain, a set of Moorish tables known as the Toledan Tables were widely used by the Jewish and Christian astronomers with minor modifications.

In the fourteenth century, the Toledan tables were surpassed by a new set of tables commonly called the Alfonsine Tables. These new tables were reputedly devised because of the encouragement of Alfonso "the Wise," King of Castile from 1273 through 1284, who wanted to have his astronomers produce a significant number of astronomical works in Spanish. The original Alfonsine Tables are lost; the earliest ones that are preserved are a form of the tables first produced in Paris around 1320. The Alfonsine Tables were produced in large quantities upon the invention of printing and were a great help to early navigators who needed exact positions of heavenly bodies to help them find their place at sea. The Tables were also used by astronomers to produce ephemerides, tabulations of planetary positions for a given year. Astronomers of the fifteenth and sixteenth centuries were not all gifted with a facility for calculation, so a number of the more talented ones produced various ephemerides to help others who did not want to take on the laborious task of calculating planetary positions from the basic astronomical tables. Regiomontanus (1436–1476) produced the first of these ephemerides for the years 1474 to 1506. Other astronomers extended them to later years. This tradition continued in later years with the English *Nautical Almanac* first appearing in 1767 and the *American Ephemeris and Nautical Almanac* in 1855.

Another way developed by ingenious astronomers to bypass the tabular calculations was through the use of graphical calculators. These generally took the form of volvelles, sets of circular papers or vellum disks attached to a backing sheet that could be rotated to obtain a readout of the information required. Volvelles first appeared in manuscript form, which could be used to determine planetary longitudes and latitudes as well as times of eclipses and dates of Easter among many other astronomical and calendrical things. They were later placed in printed books, often on separate sheets so that the owners could cut them out and mount them in the proper place in the book. The art of the volvelle reached a peak with Apianus's *Astronomicum Caesareum* (see chapter 7).

The widespread use of the Alfonsine Tables demonstrated the deficiencies of the Ptolemaic system, on which they were based. Copernicus noted that the planets' positions in the sky were not matching with their predicted postions, often by more than two degrees. Using his new model of the solar system, Copernicus produced a new table, but he did not live long enough to issue a set of handy tables as Ptolemy did. Instead, Erasmus Reinhold (1511–1553) took up that challenge and produced the Copernican-based Prutenic Tables in 1551. The Prutenic Tables rapidly supplanted the Alfonsine Tables as the standard, but they were not significantly better since most of the Prutenic Tables' data came from Ptolemaic observations. Better observational data were soon to come from the exacting Danish observer, Tycho Brahe. His data, used by the astronomer Johannes Kepler, resulted in the Rudolphine Tables of 1627. With the enormous attention to detail provided by Kepler and with his new improved planetary mechanics, the Rudolphine Tables proved to be thirty times more accurate than previous tables. An excellent example of astronomical tables can be seen in Vincenzo Renieri's (d. 1648) *Tabulae motuum caelestium universales* (Universal tables of celestial motion) of 1647. Renieri was a disciple of Galileo and succeeded the latter as professor of mathematics at Padua. The 1647 tables were an expanded version of his earlier set of 1639 and included six different tables (Alfonsine, Copernican, Prutenic, Rudolphine, and two others) in identical layouts so that the user could calculate planetary positions from any or all of them for comparison.

When Isaac Newton elucidated his laws of motion and gravitation in his *Principia* of 1687, he inaugurated a new era of study and analysis of the motions of heavenly bodies. This study became known as celestial mechanics and led to increasingly accurate determinations of the positions of planets. A major breakthrough in the development of celestial mechanics came with the application of calculus to the mathematical analysis of planetary motions. Still, celestial mechanics was fraught with difficulties because a particular planet's motion will be influenced by others as it orbits around the Sun. Nevertheless, extreme accuracy was achieved and a number of major achievements occurred in the eighteenth century: an accurate lunar theory and set of tables thanks to Alexis-Claude Clairaut (1713–1765) and Tobias Mayer (1723–1762), Clairaut's determination of the orbit of Halley's comet, and Pierre Simon, marquis de Laplace's (1749–1827) resolution of the Moon, Jupiter, and Saturn's orbits. The century concluded with Laplace's monumental multi-volume survey, the *Traité de mécanique céleste* (Treatise on celestial mechanics), published in five volumes from 1799 to 1825. Laplace treated the solar system as a problem in mathematics. All of the objects in the system were parts of a machine driven by the universal force of gravity. His work became a standard for astronomers in their efforts to solve even more difficult problems in planetary motion. *RB*

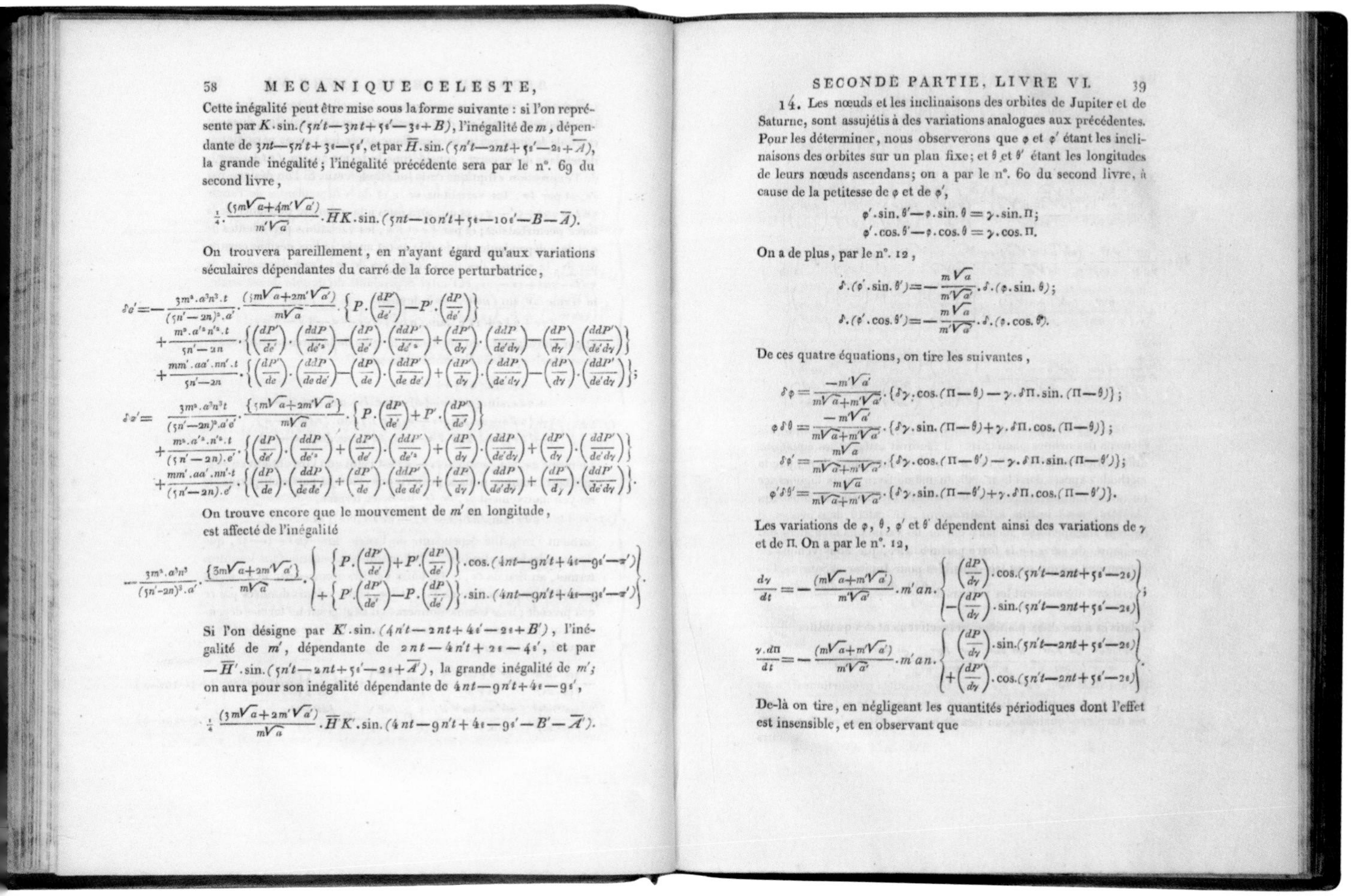
58 MECANIQUE CELESTE,

Cette inégalité peut être mise sous la forme suivante : si l'on représente par $K.\sin.(5n't-3nt+5\epsilon'-3\epsilon+B)$, l'inégalité de m, dépendante de $3nt-5n't+3\epsilon-5\epsilon'$, et par $\overline{H}.\sin.(5n't-2nt+5\epsilon'-2\epsilon+\overline{A})$, la grande inégalité ; l'inégalité précédente sera par le n°. 69 du second livre,

$$\tfrac{1}{4}.\frac{(5m\sqrt{a}+4m'\sqrt{a'})}{m'\sqrt{a}}.\overline{H}K.\sin.(5nt-10n't+5\epsilon-10\epsilon'-B-\overline{A}).$$

On trouvera pareillement, en n'ayant égard qu'aux variations séculaires dépendantes du carré de la force perturbatrice,

$$\delta e'=-\frac{3m^2.a^2n^2.t}{(5n'-2n)^2.a'}.\frac{(5m\sqrt{a}+2m'\sqrt{a'})}{m\sqrt{a}}.\left\{P.\left(\frac{dP'}{de'}\right)-P'.\left(\frac{dP}{de'}\right)\right\}$$
$$+\frac{m^2.a'^2n'^2.t}{5n'-2n}.\left\{\left(\frac{dP'}{de'}\right).\left(\frac{ddP}{de'^2}\right)-\left(\frac{dP}{de'}\right).\left(\frac{ddP'}{de'^2}\right)+\left(\frac{dP'}{d\gamma}\right).\left(\frac{ddP}{de'd\gamma}\right)-\left(\frac{dP}{d\gamma}\right).\left(\frac{ddP'}{de'd\gamma}\right)\right\}$$
$$+\frac{mm'.aa'.nn'.t}{5n'-2n}.\left\{\left(\frac{dP'}{de}\right).\left(\frac{ddP}{de\,de'}\right)-\left(\frac{dP}{de}\right).\left(\frac{ddP'}{de\,de'}\right)+\left(\frac{dP'}{d\gamma}\right).\left(\frac{ddP}{de'd\gamma}\right)-\left(\frac{dP}{d\gamma}\right).\left(\frac{ddP'}{de'd\gamma}\right)\right\};$$

$$\delta\varpi'=\frac{3m^2.a^2n^2t}{(5n'-2n)^2.a'e'}.\frac{\{5m\sqrt{a}+2m'\sqrt{a'}\}}{m\sqrt{a}}.\left\{P.\left(\frac{dP}{de'}\right)+P'.\left(\frac{dP'}{de'}\right)\right\}$$
$$+\frac{m^2.a'^2.n'^2.t}{(5n'-2n).e'}.\left\{\left(\frac{dP}{de'}\right).\left(\frac{ddP}{de'^2}\right)+\left(\frac{dP'}{de'}\right).\left(\frac{ddP'}{de'^2}\right)+\left(\frac{dP}{d\gamma}\right).\left(\frac{ddP}{de'd\gamma}\right)+\left(\frac{dP'}{d\gamma}\right).\left(\frac{ddP'}{de'd\gamma}\right)\right\}$$
$$+\frac{mm'.aa'.nn'.t}{(5n'-2n).e'}.\left\{\left(\frac{dP}{de}\right).\left(\frac{ddP}{de\,de'}\right)+\left(\frac{dP'}{de}\right).\left(\frac{ddP'}{de\,de'}\right)+\left(\frac{dP}{d\gamma}\right).\left(\frac{ddP}{de'd\gamma}\right)+\left(\frac{dP'}{d\gamma}\right).\left(\frac{ddP'}{de'd\gamma}\right)\right\}.$$

On trouve encore que le mouvement de m' en longitude, est affecté de l'inégalité

$$-\frac{3m^2.a^2n^2}{(5n'-2n)^2.a'}.\frac{\{5m\sqrt{a}+2m'\sqrt{a'}\}}{m\sqrt{a}}.\left\{\begin{array}{l}\left\{P.\left(\frac{dP'}{de'}\right)+P'.\left(\frac{dP}{de'}\right)\right\}.\cos.(4nt-9n't+4\epsilon-9\epsilon'-\varpi')\\+\left\{P'.\left(\frac{dP'}{de'}\right)-P.\left(\frac{dP}{de'}\right)\right\}.\sin.(4nt-9n't+4\epsilon-9\epsilon'-\varpi')\end{array}\right\}.$$

Si l'on désigne par $K'.\sin.(4n't-2nt+4\epsilon'-2\epsilon+B')$, l'inégalité de m', dépendante de $2nt-4n't+2\epsilon-4\epsilon'$, et par $-\overline{H}'.\sin.(5n't-2nt+5\epsilon'-2\epsilon+\overline{A}')$, la grande inégalité de m'; on aura pour son inégalité dépendante de $4nt-9n't+4\epsilon-9\epsilon'$,

$$\tfrac{1}{4}.\frac{(3m\sqrt{a}+2m'\sqrt{a'})}{m\sqrt{a}}.\overline{H}K'.\sin.(4nt-9n't+4\epsilon-9\epsilon'-B'-\overline{A}').$$

SECONDE PARTIE, LIVRE VI. 39

14. Les nœuds et les inclinaisons des orbites de Jupiter et de Saturne, sont assujétis à des variations analogues aux précédentes. Pour les déterminer, nous observerons que φ et φ' étant les inclinaisons des orbites sur un plan fixe; et θ et θ' étant les longitudes de leurs nœuds ascendans; on a par le n°. 60 du second livre, à cause de la petitesse de φ et de φ',

$$\varphi'.\sin.\theta'-\varphi.\sin.\theta=\gamma.\sin.\Pi;$$
$$\varphi'.\cos.\theta'-\varphi.\cos.\theta=\gamma.\cos.\Pi.$$

On a de plus, par le n°. 12,

$$\delta.(\varphi'.\sin.\theta')=-\frac{m\sqrt{a}}{m'\sqrt{a'}}.\delta.(\varphi.\sin.\theta);$$
$$\delta.(\varphi'.\cos.\theta')=-\frac{m\sqrt{a}}{m'\sqrt{a'}}.\delta.(\varphi.\cos.\theta).$$

De ces quatre équations, on tire les suivantes,

$$\delta\varphi=\frac{-m'\sqrt{a'}}{m\sqrt{a}+m'\sqrt{a'}}.\{\delta\gamma.\cos.(\Pi-\theta)-\gamma.\delta\Pi.\sin.(\Pi-\theta)\};$$
$$\varphi\delta\theta=\frac{-m'\sqrt{a'}}{m\sqrt{a}+m'\sqrt{a'}}.\{\delta\gamma.\sin.(\Pi-\theta)+\gamma.\delta\Pi.\cos.(\Pi-\theta)\};$$
$$\delta\varphi'=\frac{m\sqrt{a}}{m\sqrt{a}+m'\sqrt{a'}}.\{\delta\gamma.\cos.(\Pi-\theta')-\gamma.\delta\Pi.\sin.(\Pi-\theta')\};$$
$$\varphi'\delta\theta'=\frac{m\sqrt{a}}{m\sqrt{a}+m'\sqrt{a'}}.\{\delta\gamma.\sin.(\Pi-\theta')+\gamma.\delta\Pi.\cos.(\Pi-\theta')\}.$$

Les variations de φ, θ, φ' et θ' dépendent ainsi des variations de γ et de Π. On a par le n°. 12,

$$\frac{d\gamma}{dt}=-\frac{(m\sqrt{a}+m'\sqrt{a'})}{m'\sqrt{a'}}.m'an.\left\{\begin{array}{l}\left(\frac{dP}{d\gamma}\right).\cos.(5n't-2nt+5\epsilon'-2\epsilon)\\-\left(\frac{dP'}{d\gamma}\right).\sin.(5n't-2nt+5\epsilon'-2\epsilon)\end{array}\right\};$$
$$\frac{\gamma.d\Pi}{dt}=-\frac{(m\sqrt{a}+m'\sqrt{a'})}{m'\sqrt{a'}}.m'an.\left\{\begin{array}{l}\left(\frac{dP}{d\gamma}\right).\sin.(5n't-2nt+5\epsilon'-2\epsilon)\\+\left(\frac{dP'}{d\gamma}\right).\cos.(5n't-2nt+5\epsilon'-2\epsilon)\end{array}\right\}$$

De-là on tire, en négligeant les quantités périodiques dont l'effet est insensible, et en observant que

Pierre Simon Laplace. *Traité de mécanique céleste* (Paris : De l'Imprimerie de Crapelet, 1802). The dense mathematical formulas found in works like Laplace's indicate the high level of analysis applied to celestial mechanics at the end of the eighteenth century. (Huntington Library: RB 478637, v. 3).

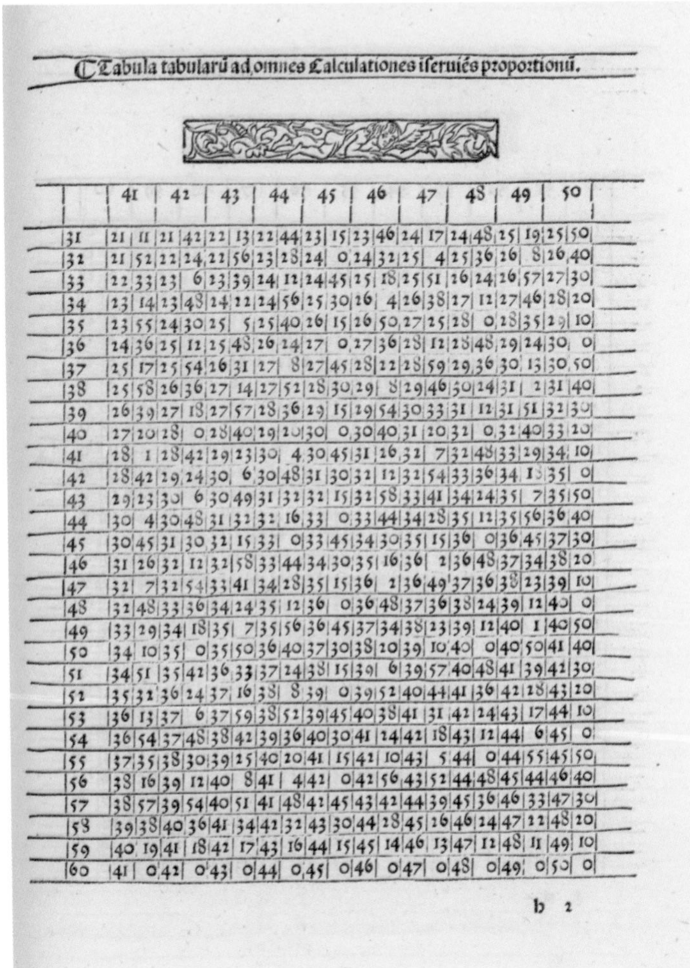

Tabula tabularũ ad omnes Calculationes iseruiẽs proportionũ.

	41		42		43		44		45		46		47		48		49		50	
31	21	11	21	42	22	13	22	44	23	15	23	46	24	17	24	48	25	19	25	50
32	21	52	22	24	22	56	23	28	24	0	24	32	25	4	25	36	26	8	26	40
33	22	33	23	6	23	39	24	12	24	45	25	18	25	51	26	24	26	57	27	30
34	23	14	23	48	24	22	24	56	25	30	26	4	26	38	27	12	27	46	28	20
35	23	55	24	30	25	5	25	40	26	15	26	50	27	25	28	0	28	35	29	10
36	24	36	25	12	25	48	26	24	27	0	27	36	28	12	28	48	29	24	30	0
37	25	17	25	54	26	31	27	8	27	45	28	22	28	59	29	36	30	13	30	50
38	25	58	26	36	27	14	27	52	28	30	29	8	29	46	30	24	31	2	31	40
39	26	39	27	18	27	57	28	36	29	15	29	54	30	33	31	12	31	51	32	30
40	27	20	28	0	28	40	29	20	30	0	30	40	31	20	32	0	32	40	33	20
41	28	1	28	42	29	23	30	4	30	45	31	26	32	7	32	48	33	29	34	10
42	28	42	29	24	30	6	30	48	31	30	32	12	32	54	33	36	34	18	35	0
43	29	23	30	6	30	49	31	32	32	15	32	58	33	41	34	24	35	7	35	50
44	30	4	30	48	31	32	32	16	33	0	33	44	34	28	35	12	35	56	36	40
45	30	45	31	30	32	15	33	0	33	45	34	30	35	15	36	0	36	45	37	30
46	31	26	32	12	32	58	33	44	34	30	35	16	36	2	36	48	37	34	38	20
47	32	7	32	54	33	41	34	28	35	15	36	2	36	49	37	36	38	23	39	10
48	32	48	33	36	34	24	35	12	36	0	36	48	37	36	38	24	39	12	40	0
49	33	29	34	18	35	7	35	56	36	45	37	34	38	23	39	12	40	1	40	50
50	34	10	35	0	35	50	36	40	37	30	38	20	39	10	40	0	40	50	41	40
51	34	51	35	42	36	33	37	24	38	15	39	6	39	57	40	48	41	39	42	30
52	35	32	36	24	37	16	38	8	39	0	39	52	40	44	41	36	42	28	43	20
53	36	13	37	6	37	59	38	52	39	45	40	38	41	31	42	24	43	17	44	10
54	36	54	37	48	38	42	39	36	40	30	41	24	42	18	43	12	44	6	45	0
55	37	35	38	30	39	25	40	20	41	15	42	10	43	5	44	0	44	55	45	50
56	38	16	39	12	40	8	41	4	42	0	42	56	43	52	44	48	45	44	46	40
57	38	57	39	54	40	51	41	48	42	45	43	42	44	39	45	36	46	33	47	30
58	39	38	40	36	41	34	42	32	43	30	44	28	45	26	46	24	47	22	48	20
59	40	19	41	18	42	17	43	16	44	15	45	14	46	13	47	12	48	11	49	10
60	41	0	42	0	43	0	44	0	45	0	46	0	47	0	48	0	49	0	50	0

b 2

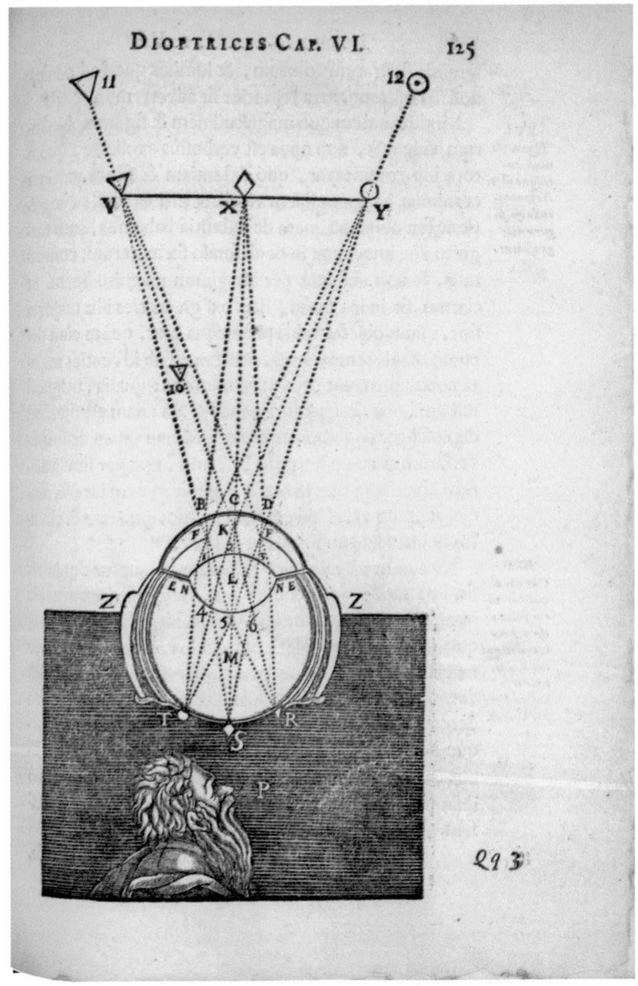

Alfonso X el sabio. *Tabulae astronomicae* (Venice: Per Io. Antonium et fratres de Sabio, 1492).One of the many pages of tables from the Alphonsine Tables to be used by astronomers determining planetary positions. (Huntington Library: RB 89420)

René Descartes. *Principia philosophiae* (Amsterdam: Elzevir, 1644). Descartes was an expert in optics, and here he illustrates the optical path that light follows as it enters the human eye. (Huntington Library: RB 336030)

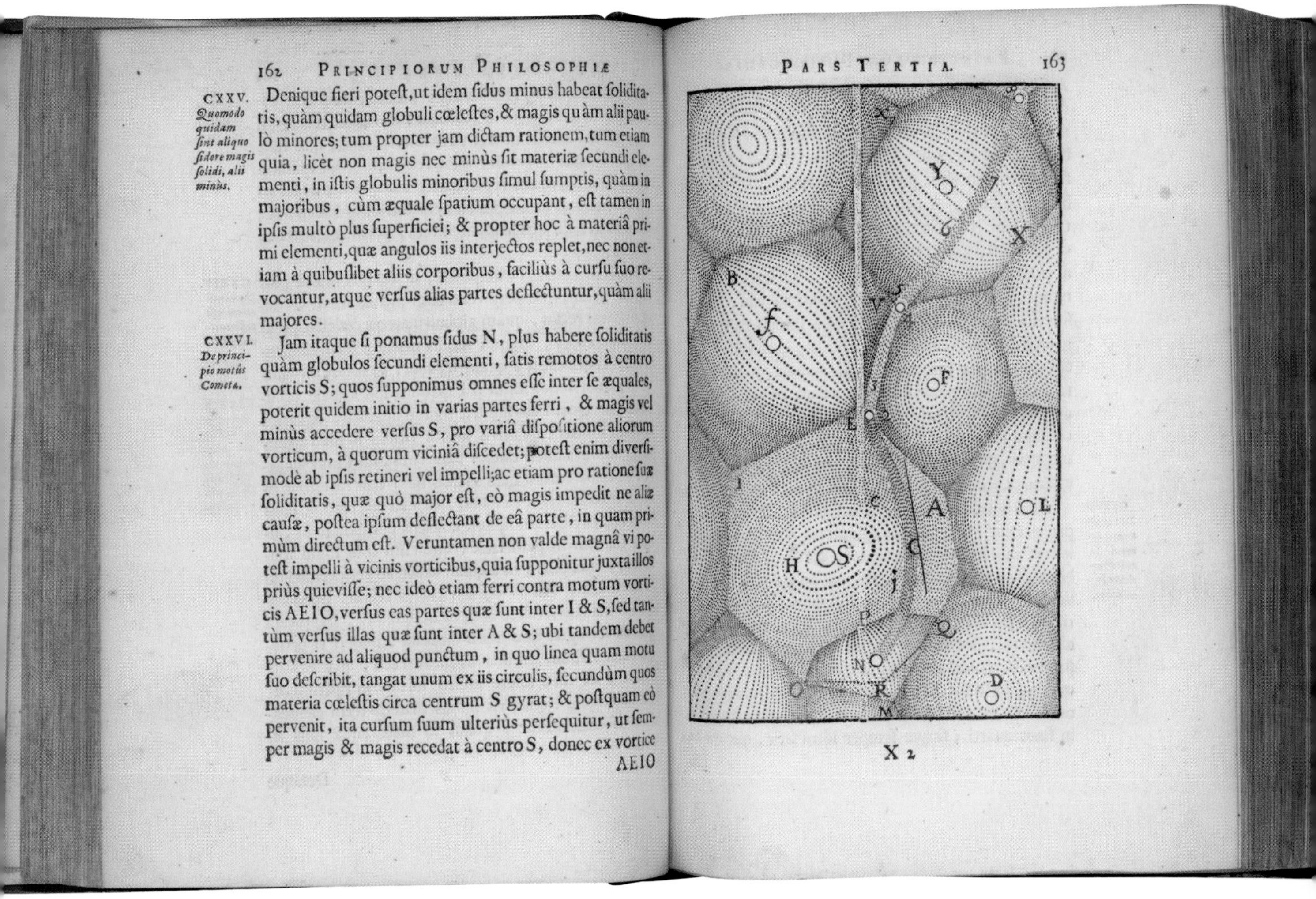

162 PRINCIPIORUM PHILOSOPHIÆ

CXXV. *Quomodo quidam ſint aliquo ſidere magis ſolidi, alii minus.*

Denique fieri poteſt, ut idem ſidus minus habeat ſoliditatis, quàm quidam globuli cœleſtes, & magis quàm alii paulò minores; tum propter jam dictam rationem, tum etiam quia, licèt non magis nec minùs ſit materiæ ſecundi elementi, in iſtis globulis minoribus ſimul ſumptis, quàm in majoribus, cùm æquale ſpatium occupant, eſt tamen in ipſis multò plus ſuperficiei; & propter hoc à materiâ primi elementi, quæ angulos iis interjectos replet, nec non etiam à quibuſlibet aliis corporibus, faciliùs à curſu ſuo revocantur, atque verſus alias partes deflectuntur, quàm alii majores.

CXXVI. *De principio motûs Cometæ.*

Jam itaque ſi ponamus ſidus N, plus habere ſoliditatis quàm globulos ſecundi elementi, ſatis remotos à centro vorticis S; quos ſupponimus omnes eſſe inter ſe æquales, poterit quidem initio in varias partes ferri, & magis vel minùs accedere verſus S, pro variâ diſpoſitione aliorum vorticum, à quorum viciniâ diſcedet; poteſt enim diverſimodè ab ipſis retineri vel impelli; ac etiam pro ratione ſuæ ſoliditatis, quæ quò major eſt, eò magis impedit ne aliæ cauſæ, poſtea ipſum deflectant de eâ parte, in quam primùm directum eſt. Veruntamen non valde magnâ vi poteſt impelli à vicinis vorticibus, quia ſupponitur juxta illos priùs quieviſſe; nec ideò etiam ferri contra motum vorticis AEIO, verſus eas partes quæ ſunt inter I & S, ſed tantùm verſus illas quæ ſunt inter A & S; ubi tandem debet pervenire ad aliquod punctum, in quo linea quam motu ſuo deſcribit, tangat unum ex iis circulis, ſecundùm quos materia cœleſtis circa centrum S gyrat; & poſtquam eò pervenit, ita curſum ſuum ulteriùs perſequitur, ut ſemper magis & magis recedat à centro S, donec ex vortice

AEIO

PARS TERTIA. 163

X 2

Descartes's conception of the universe had each star surrounded by vortices in the aether. This illustration shows his explanation for how a comet would travel from one vortex to the next as it moves through the universe.

CANCER
GEMINI
TAVRVS
ARIES
PISCES
AQVARIVS
CAPRICORNVS
SAGITTARIVS
SCORPIVS
LIBRA
VIRGO
LEO
CAPVT
CAVDA
LATITVDO LVNAE MERIDIONALIS
LATITVDO LVNAE SEPTENTRIONALIS
VMBRA TERRAE

F III

SEVEN PETRUS APIANUS AND THE *ASTRONOMICUM CAESAREUM*

Petrus Apianus. *Astronomicum Caesareum* (Ingolstadt: From the author's press, 1540). This dragon image perfectly captures the spirit of boldness and artistry evident throughout this work. (Huntington Library: RB 32891)

THE MOST INTRICATE and lavishly designed example of the bookmaker's art in the sixteenth century, and perhaps the most beautiful scientific book ever printed, was the *Astronomicum Caesareum* of Petrus Apianus (1495–1552), also known as Peter Bienewitz and Peter Apian.[3] This great landmark of printing was completed in 1540 at Apianus's private press at Ingolstadt, along the Danube in Germany. Produced for Charles V and his brother Ferdinand of Spain, the volume was carefully planned and beautifully executed. Its pages were large, brilliantly hand-colored, and filled with carefully placed *volvelles* (paper disks) in as many as six layers. Volvelles, which are rotating paper devices used to perform calculations for solving problems of the calendar or the positions of the Sun, Moon, and other planets, were not uncommon in sixteenth-century books, and trace their origins back to earlier manuscripts. Volvelles had been used before in books but never on this scale and in this number; they were always aids to the text, rather than vice versa as with the *Astronomicum*. The volvelles replaced a large amount of tabular data, and proved to be very accurate—yielding positions within one degree of the more accurate, but more labor-intensive, columns of numbers they replaced. In their fully colored and elegant glory, the volvelles helped to "graphically display Ptolemaic astronomy in a fashion fit for a monarch's eyes," as Owen Gingerich noted in 1971.[4] Most copies of the work were printed on unusually thick paper—used for its intrinsic beauty and weight but also to support the mechanisms of the volvelles.

This work, which took Apianus eight years to produce, is really a scientific calculating instrument as much as a book. Copies of this spectacular book have slowly come to light. By 1901 thirty-three copies had been located; several more turned up in a 1933 survey. The most comprehensive survey yet by far, undertaken by Gingerich and published in 1997, finds 111 copies extant, with the largest number in Germany (34), the second-largest in the United States (17), and thirteen copies each in France and Italy.

Beyond its clever design and beautiful appearance, the *Astronomicum* is notable for Apianus's pioneer observations of comets (he describes the appearances of five comets, including Halley's) and his statement that comets point their tails away from the Sun—a dramatic observation of the phenomenon by which either the "solar wind" of charged particles or the sunlight itself creates pressure radiating outward from the Sun. He was also acutely aware of the challenges faced by navigators and explorers in determining their bearings and suggested how solar eclipses might be used to measure longitude, one of the most pressing problems of the day.

Additionally, his circular representation of the cosmos—shown on a deep blue background in the Huntington's copy—includes the first European rendition of the constellation Ursa Major—the Great Bear—the faint companion of the brighter Ursa Minor. Also known as "Alcor," Ursa Major was not cataloged by Ptolemy. Islamic astronomers who did see the star often referred to it as "the overlooked star." Ursa Minor has also been represented in Islamic views of the constellation as a team of horses pulling a wagon.

Before creating his masterwork, Apianus studied mathematics and astronomy at Leipzig and Vienna, and quickly established a reputation as an outstanding mathematician and geographer. His first major work, the *Cosmographia* (Description of the Universe) appeared in 1524. Using an ingenious and simple diagram, he defined terrestrial grids, described the use of maps and simple surveying, and provided thumbnail sketches of the continents; this volume also included three volvelles. The *Cosmographia* was modified in 1533 with the addition of an important chapter by Gemma Frisius (1508–1555), a physician and cartographer, who proposed and illustrated for the first time the principle of triangulation as a means of accurately locating and mapping different regions. The *Cosmographia* became one of the most popular texts of its day; it was translated into all major European languages and was issued in no fewer than forty-five editions by at least eighteen publisher/printers.

The success of this and earlier works led to Apianus's appointment as professor of mathematics at the University of Ingolstadt, where he remained until his death. Apianus and Charles V had a long and fruitful relationship, dating back to least the early 1530s, when Apianus was granted an imperial privilege. Apianus gained patronage and privilege, and Charles had the attentions and instruction of a talented astronomer and printer. Following the publication of the *Astronomicum Caesareum*, Apianus was appointed court mathematician to Charles V and was knighted, along with his three brothers. In the following years, the success of Apianus's books and his favor with the emperor brought him a substantial degree of wealth, prestige, and power. In 1544 he was granted special legal privileges, including the authority to legitimize illegitimate children and to grant higher degrees. Throughout his life, Apianus was not just a publisher but also an exacting scholar involved in mathematical publishing, cartography, and instrument making. His instrument designs included a new type of surveying quadrant and armillary spheres. *DL*

OPPOSITE

Petrus Apianus. *Astronomicum Caesareum* (Ingolstadt: From the author's press, 1540). This circular image of the cosmos includes the first European representation of the constellation Ursa Major, the Great Bear. (Huntington Library: RB 32891)

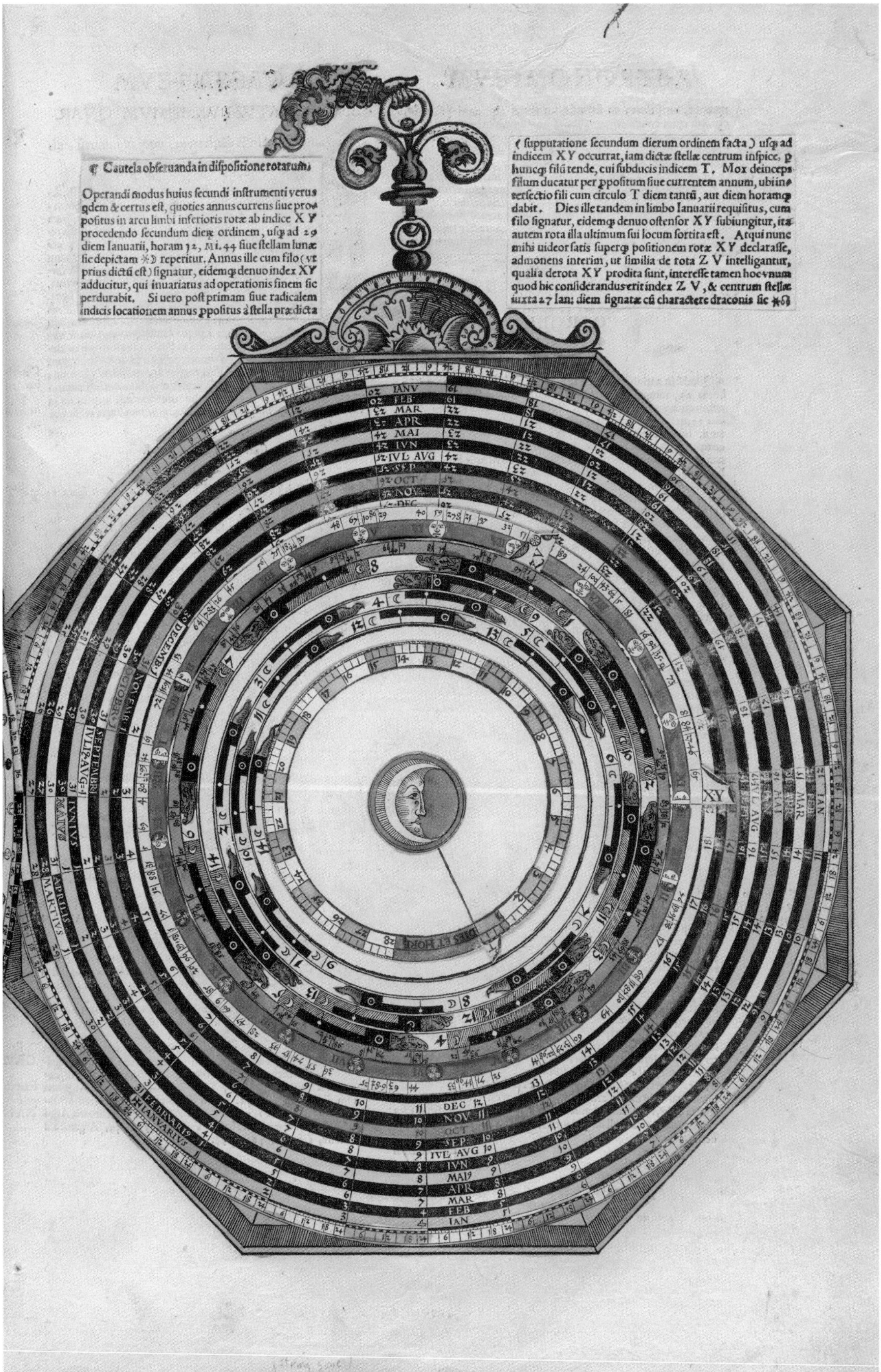

Petrus Apianus. *Astronomicum Caesareum* (Ingolstadt: From the author's press, 1540).

THIS PAGE: The actual design of charts to plot celestial data strongly echoes the Ptolemaic view of the universe as consisting of perfect circles. This illustration, for instance, is strikingly similar to that of the view of the seventh day of creation shown on page 48 of this book.

OPPOSITE: The volvelles in this work allowed the user to calculate planetary positions plus a variety of calendrical and astrological data. They yield positions always within one degree of the more accurate (but more tedious) numerical tables that they replaced. Technically complex, they consist of as many as six disks, some hidden behind assemblages of other disks in order to allow for movements around multiple axes. (Huntington Library: RB 32891)

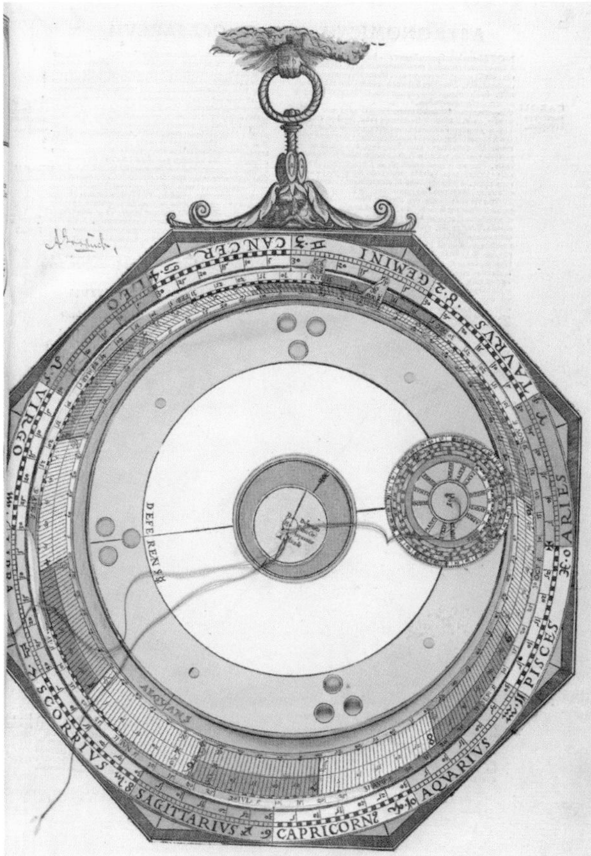
CANCER
GEMINI
TAVRVS
LEO
VIRGO
ARIES
PISCES
AQVARIVS
CAPRICORN
SAGITTARIVS
SCORPIVS
DEFERENS

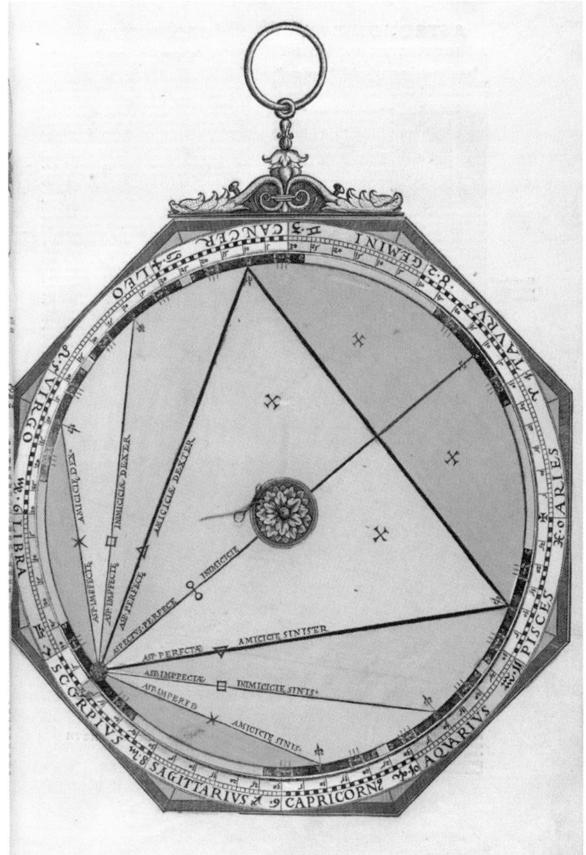
CANCER
GEMINI
TAVRVS
LEO
VIRGO
LIBRA
ARIES
PISCES
AQVARIVS
CAPRICORN
SAGITTARIVS
SCORPIVS
AMICICIE DEXTER
INIMICICIE DEXTER
AMICICIE SINISTER

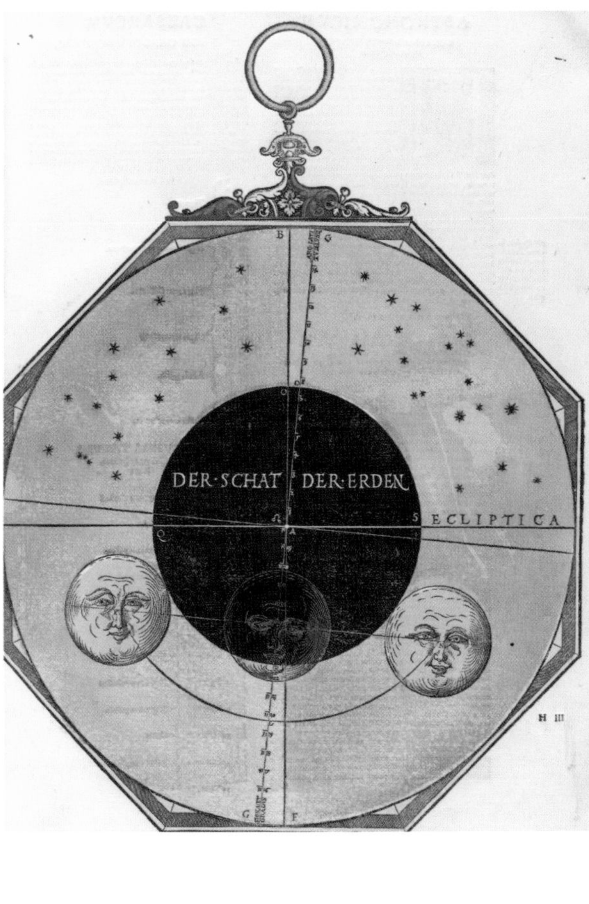
DER·SCHAT DER·ERDEN
ECLIPTICA

VIA·LVNAE
ECLIPTICA
VMBRA TERRE
ECLIPTICA
VIA·LVNAE

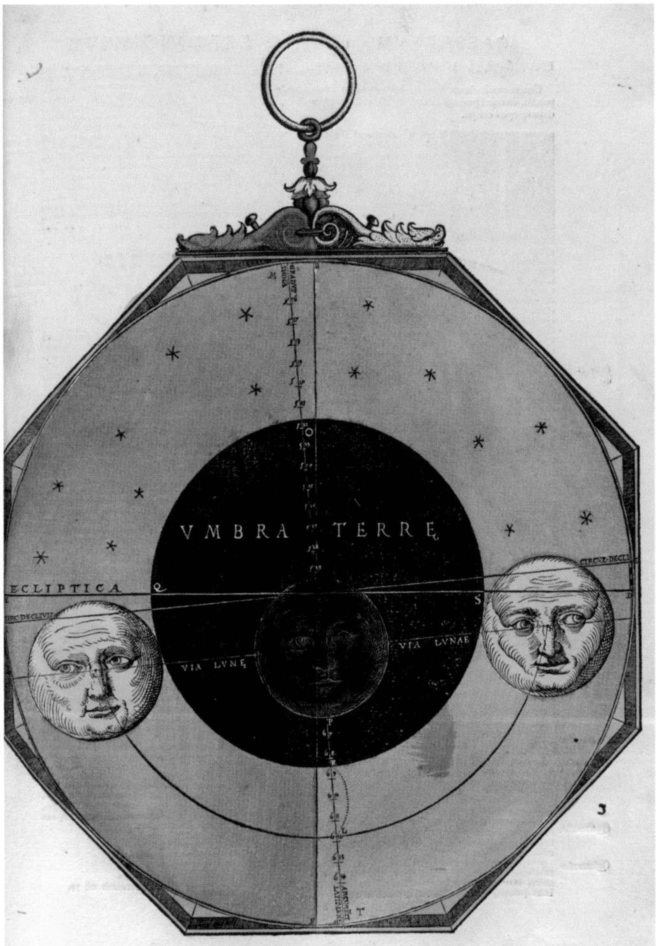

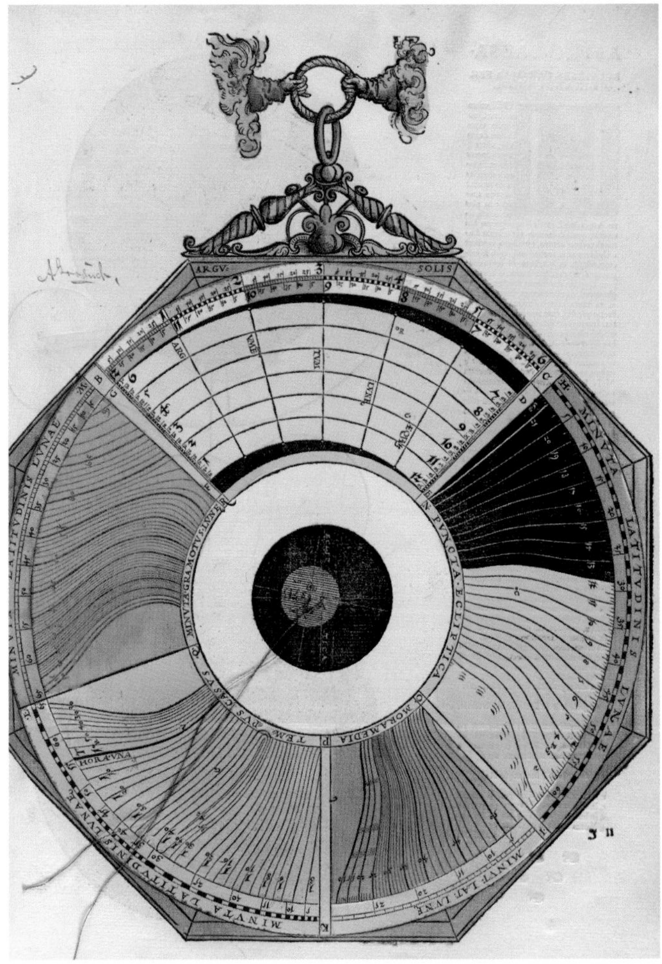

Petrus Apianus. *Astronomicum Caesareum* (Ingolstadt: From the author's press, 1540).

ABOVE: These volvelles allowed the plotting and tracking of lunar motion and latitude. As the most visible element of the night sky, the Moon is represented in various forms in the Apianus atlas, including information on lunar eclipses.

OPPOSITE: The rotating paper disks in the work were all hand-colored in Apianus's printing shop (and not by the buyers themselves, as was often the case for sixteenth-century books). The sheets were colored before they were cut. Of the approximately 120 copies known, only a single uncolored example exists—in the Staatsbibliothek in Munich. (Huntington Library: RB 32891)

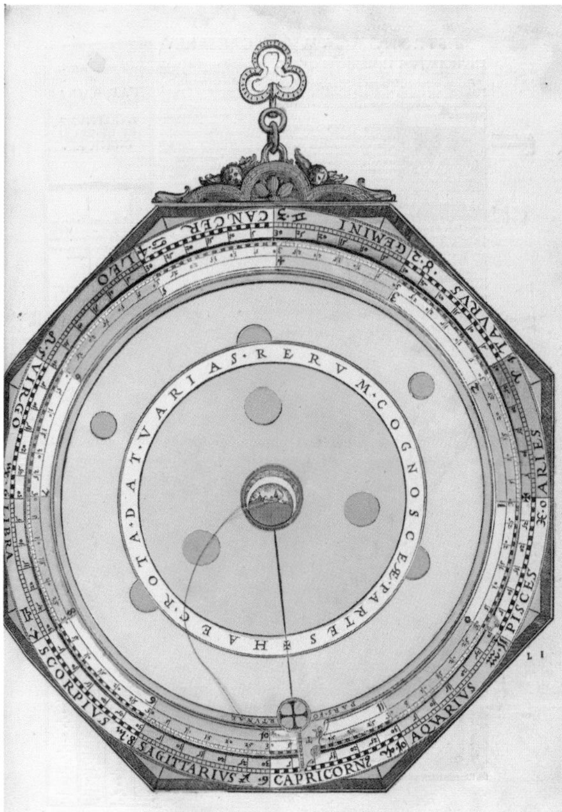
CANCER
GEMINI
TAVRVS
ARIES
PISCES
AQVARIVS
CAPRICORN
SAGITTARIVS
SCORPIVS
LIBRA
VIRGO
LEO
HAEC ROTA DAT VARIAS RERVM COGNOSCERE PARTES
L I
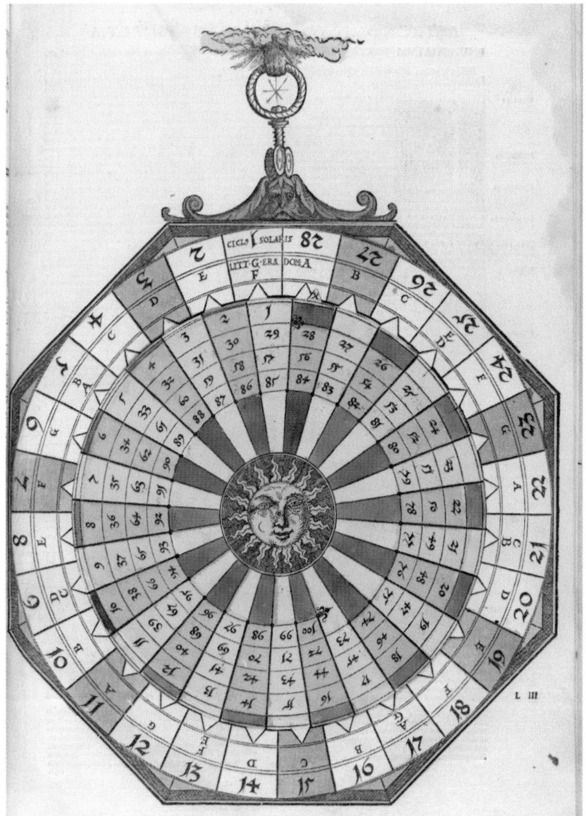
CICLVS SOLARIS
L III
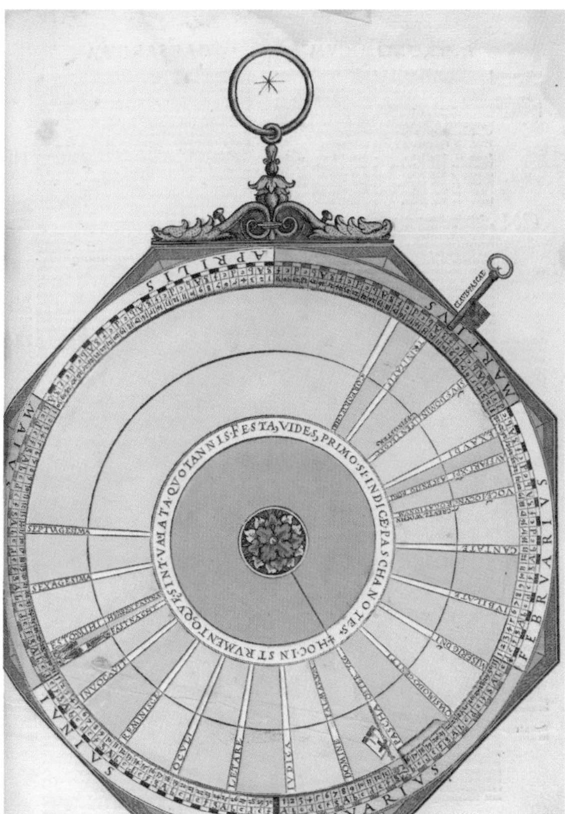
APRILIS
MARTIVS
FEBRVARIVS
IANVARIVS
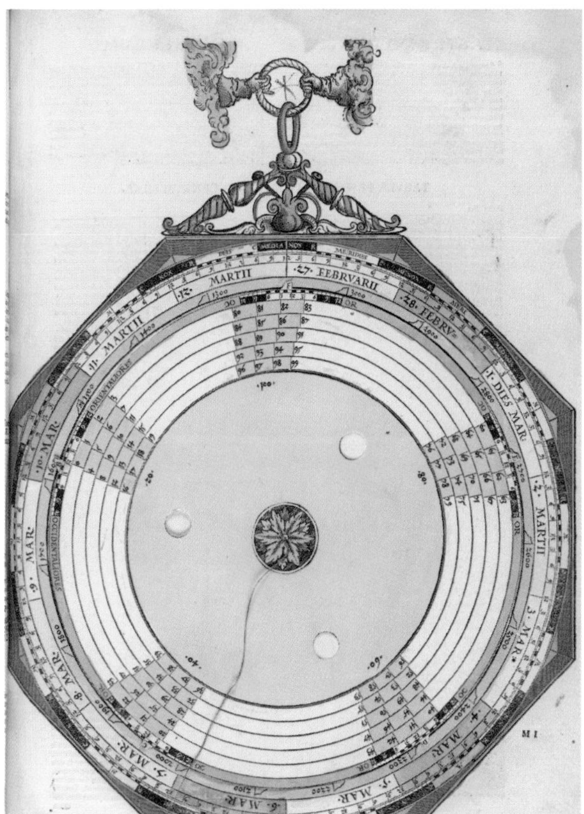
MARTII
FEBRVARII
M I

II

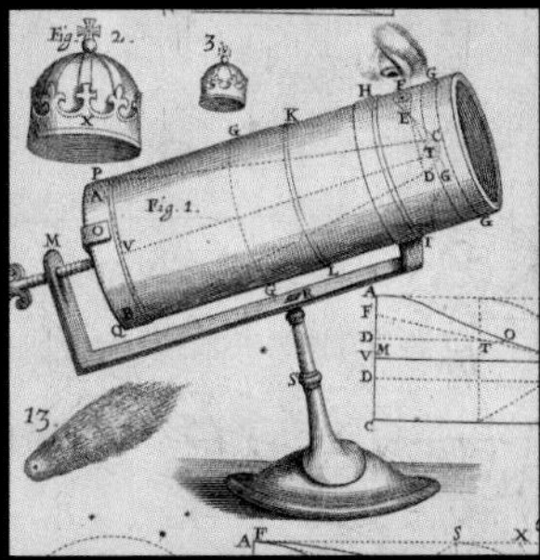

TECHNOLOGI

OF

OBSERVING

TRAITÉ DE COSMOGRAPHIE.

LIVRE PREMIER.

De la Sphere du Monde.

Définitions nécessaires à ce Traité.

1. LA Sphere que l'on appelle aussi Globe ou Boule, est une figure solide comprise d'une seule superficie courbe, en laquelle toutes les lignes droites menées du centre à la superficie, sont égales entr'elles.

2. Le centre de la Sphere est ce même point duquel toutes les lignes tirées à la superficie sont égales entr'elles.

3. Le diametre de la Sphere est une ligne droite qui passe par le centre, & se termine de part & d'autre à la superficie.

4. L'axe ou l'essieu de la Sphere est l'un de ses diametres sur lequel elle tourne.

EXPLICATION.

Si ayant percé une orange avec une longue éguille, laquelle passe par le milieu, on la fait tourner autour de cette éguille, elle pourra être nommée son axe.

5. Les poles de la Sphere sont deux points opposés en sa superficie, & qui sont à l'extrémité de l'axe.

EIGHT | MINIATURE SPINNING WORLDS
Celestial Globes and Armillaries

THE SPHERICAL REPRESENTATIONS of the heavens that astronomers put to practical use for many centuries continue to provide aesthetic satisfactions. They allowed astronomers to view the skies in miniature; to demonstrate the ways they believed the stars and planets worked; and sometimes, to make accurate observations using large versions, as Tycho Brahe did in his observatory at Uraniborg. They were also often highly artful creations, carefully and beautifully constructed, that were used as furnishings and works of art.[5]

The word *armillary* comes from the Latin word *armilla*, meaning "bracelet." Armillary spheres are supported by a fixed ring representing the horizon of the observer and could be adapted to the latitude and longitude of the country where they were being used. Composed of two or more nested rings, armillaries are usually of two types: observational and demonstrational. Most observational armillaries, such as Tycho's, were large—up to three meters in diameter—and constructed from brass, while demonstrational armillaries were smaller, and often made of wood and pasteboard instead of metal.

Armillaries date to the time of the Greek astronomer Eratosthenes, in the 3rd century B.C.E. Representations of the celestial sphere, consisting of graduated metallic rings representing the equator, the ecliptic and certain meridians and parallels, were widely used by natural philosophers of the Alexandrian school. The observational armillary was also developed independently in China, where it served as a very important astronomical tool.

Early armillaries showed a geocentric universe, with the Earth at the center of the armillary and rings carrying images of the Moon, Sun, and planets revolving around it. The observational versions of these tools enabled astronomers to find the coordinates of stars. After Copernicus's persuasive heliocentric argument caught on, the Sun moved to the center of models of armillary spheres, accompanied by—in the more complicated models—the Moon and planets. By correctly placing bodies inside the sphere and the rings, it was possible to solve problems of spherical astronomy and compute the coordinates of the stars on the celestial sphere. The dimensions of such instruments made them lack precision and so they were used for more didactic and explicative purposes. But large spheres did exist. Armillaries ranged in size from the very small—such works as the very small fifteen-centimeter diameter sphere built in 1720 by the Londoner John Rowley (d. 1728)—up to the three-meter version made in wood, engraved and gilded by Antonio Santucci (active 1590–1619) at the end of the sixteenth century, that is now on display at the Instituto e Museo di Storia della Scienza in Florence.

The late seventeenth and early eighteenth centuries were the apogee of armillary

PRECEDING SPREAD

The massive forty-foot reflecting telescope designed by British astronomer William Herschel in the late eighteenth century (Huntington Library: RB 98681. v.85)

OPPOSITE

Nicolas Bion. *L'usage des globes...* (6th edition. Paris: Jacques Guerin, 1751). Cosmography, or the study of the heavens, was a highly refined and fully formulated science from the sixteenth century forward. In the broadest sense, it was defined as the study of the universe; celestial globes served as models for the study of the universe. (Huntington Library: RB 487000.47)

sphere production and beautiful examples were built, often gilded with spectacular engravings of zodiacal signs and important positions on the different rings. Their style and elegance doomed them, however, as other means of observing the heavens, primarily advances in the design of the telescope, became more effective tools. Armillaries became objects of fashion more than of study. At the end of the eighteenth century, especially in France, armillary spheres were mass-produced, primarily in wood covered with colored paper.

The celestial globe has held an even more important place in the history of astronomy than armillary spheres. It competed, not too successfully, with the flat map, which was more significant for scientific use because it could be more easily updated and reprinted, created on a larger scale, and readily marked up and even held up against the night sky. But one of the great benefits of celestial globes is that they were—and are—capable of representing the heavens without the distortions inherent in using a flat surface to represent a universe visible in 360 degrees around a spherical Earth.

Nuremberg was the first center of both terrestrial and celestial globe production in Europe, just as it was the first center for the production of scientific books. Johann Schöner (1477–1574) created his workshop to mass-produce globes there in the early sixteenth century. By the close of the sixteenth century, however, Amsterdam had become the new center for production and international distribution of the increasingly popular and available globes of Earth and heaven. Jodocus Hondius (1563–1612) and William Blaeu (1571–1638) were among the principal globe-makers there, and fierce competitors with different motivations. Whereas Hondius badly needed the money his growing globe business produced, Blaeu was more financially stable and, as the son of a merchant working in the counting-house of a relative, his future in commerce was assured. But his interests in science eclipsed any possible career in finance he might have followed. Blaeu was encouraged by Adriaen Anthonisz, an engineer in the Netherlands, to make a celestial globe as a practical aid to astronomy. He studied with the famous Danish astronomer Tycho Brahe to obtain the proper scientific training. Soon after the completion of his celestial globe. Blaeu brought out a terrestrial globe.

In 1603, at the end of the second phase of globe production in Amsterdam, both globe manufacturers had produced pairs of globes—one celestial and one terrestrial sphere in each set—in several sizes: large globes of at least a foot in diameter; small, 8 to 9 inches; as well as miniatures. Both the terrestrial and celestial globes included the latest discoveries. The celestial globes were in fact drawn in a completely new manner. The production of globes developed rapidly, and resulted in an astonishing diversity of different varieties. Globe making in Amsterdam soon dominated world production. Between 1605 and 1612 both Hondius and Blaeu extended their cartographic collection. However, they developed in different directions. Hondius threw himself into the production of atlases after he managed to purchase the copperplates for printing the *Atlas* of mapmaker Gerardus Mercator (1512–1594), after whom the Mercator map projection—still in use—was named following its creation in 1569. Blaeu, on the other hand, specialized in printing individual unbound maps, publishing charts and pilot-guides, and in making instruments for navigation.

One important writer detailing the evolution of armillary and celestial globes was the astronomer, mathematician and cartographer Egnatio Danti (1537–1586), who wrote *Dell'uso et fabrica dell' astrolabio et del plansiferio* (Of the use and fabrication of the astrolabe and the planisphere), first published in Italy in 1578. Danti performed

4 PARTE PRIMA DEL

DELLA SFERA DI PROCLO.

M.A.N Circulo che rappreſenta il Meridiano.
N.M. L'Orizonte.
B.D.O.C. il Coluro ſolſtiziale.
A.B. il Coluro Equinoziale, & l'aſſe del Mondo.
O.R. il circulo Artico, che ſta tutto ſopra l'Orizonte.
E.F. il Tropico Eſtiuo del Cancro.
D.C. L'Equinoziale.
H.G. il Tropico di Bruma.
K.I. il circulo antartico, che ſta tutto ſotto l'Orizonte
F.A. il Zodiaco diuiſo ne' 12 ſegni.
A. il Polo Artico.
B. il Polo Antartico.

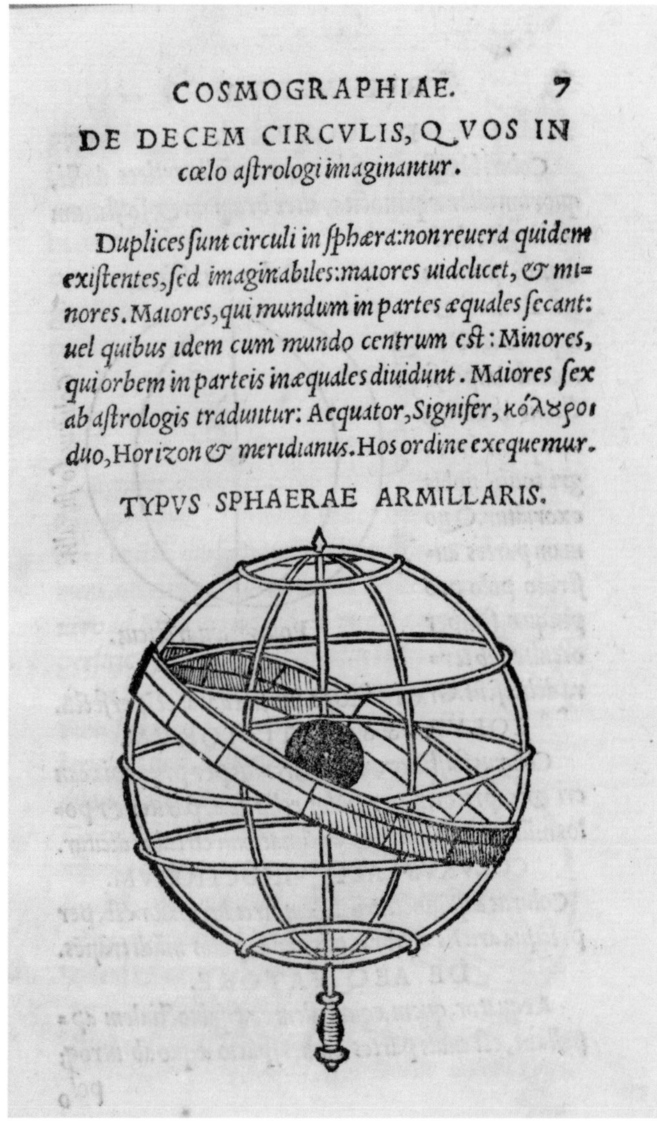

COSMOGRAPHIAE. 7

DE DECEM CIRCVLIS, QVOS IN cœlo aſtrologi imaginantur.

Duplices ſunt circuli in ſphæra: non reuera quidem exiſtentes, ſed imaginabiles: maiores uidelicet, & minores. Maiores, qui mundum in partes æquales ſecant: uel quibus idem cum mundo centrum eſt: Minores, qui orbem in parteis inæquales diuidunt. Maiores ſex ab aſtrologis traduntur: Aequator, Signifer, κόλουροι duo, Horizon & meridianus. Hos ordine exequemur.

TYPVS SPHAERAE ARMILLARIS.

ABOVE LEFT

Egnatio Danti. *Dell'uso et fabrica dell' astrolabio et del plansiferio...* (Florence: Appresso i Giunti, 1578). The Italian mapmaker Danti detailed the creation and use of the astrolabe and the planisphere, two other early astronomical instruments. He was also interested in globes and their uses. The equatorial armillary sphere shown here is labeled to show the North and South poles, the Arctic Circle, the equinox, and other aspects. (Huntington Library: RB 487000.95)

ABOVE RIGHT

Petrus Apianus. *Cosmographiae Introducto* (3rd edition. Ingolstadt: From the author's press, 1533). The armillary sphere provided a decorative element as much as an astronomical one; here, it function as a printed *objet d'art*; the bands contain no numbers. (Huntington Library: RB 13808)

vital work as a scientist—as one of the earliest astronomers to calculate just how far out of date the calendar had become before the use of the leap year, he was one of the most important figures in the several decades of calendar reform that took place in the first half of the sixteenth century with the transition from the Julian to the Gregorian calendar. He published his grandfather's translation of *Tractatus de Sphaera* (Treatise on the spheres), an important astronomical work written by the scholar Johannes de Sacrobosco (1195–1256). Danti included his own commentary with the translation and also published other astronomical work and mathematical works. He himself translated Ptolemy's *Geography* into Italian, prepared huge mural maps in Florence (in the Palazzo Vecchio) and later in the Vatican, and published a work on a surveying instrument that he improved upon. But he is probably best known for his work on the armillary sphere and globe. To ascertain that the calendar was out of date, he constructed an astronomical quadrant and an armillary for observations designed to determine the true equinox in order to correct the calendar. He both worked extensively with these astronomical tools and published the aforementioned book and others on armillary spheres and other astronomical instruments. *DL*

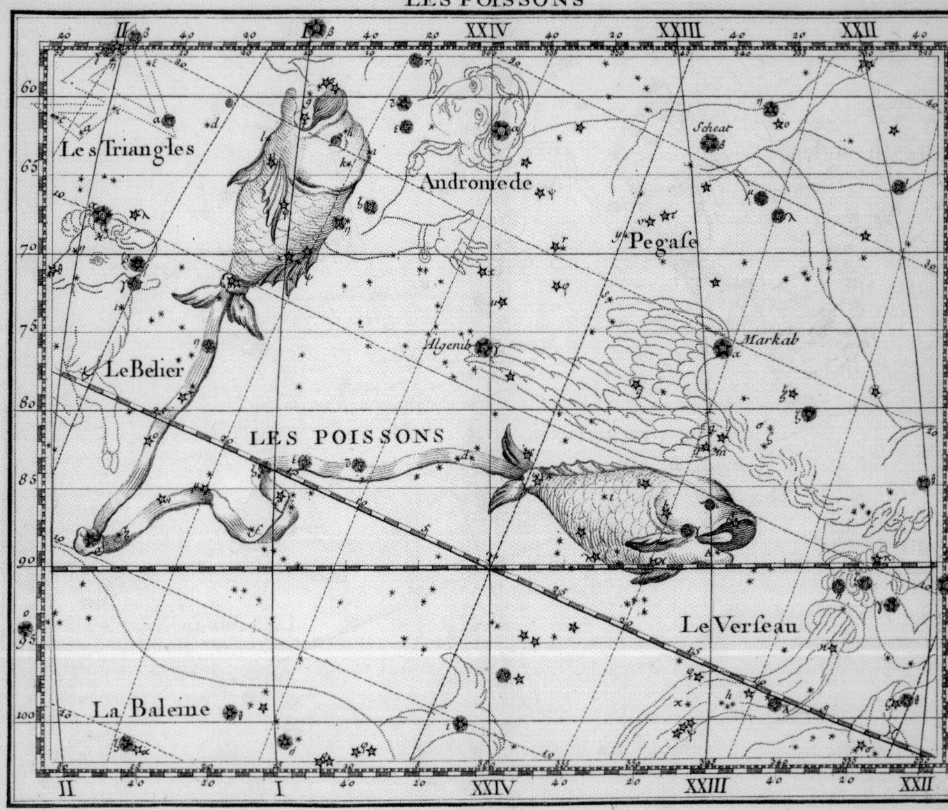
Les Triangles
Andromede
Pegase
Scheat
Markab
Algenib
Le Belier
LES POISSONS
Le Verseau
La Baleine

NINE | DESIGNING THE HEAVENS: *Constellations and Star Charts*

CELESTIAL CARTOGRAPHY has a long, complex, and fascinating history.[6] It has involved artistry (such as the work of Albrecht Dürer and Johann Bayer), vanity and flattery (cartographers would often include images of their patrons as figures in the sky in attempts to confer celestial immortality on their leaders), and almost every conceivable exotic object or beast (including the bird of paradise, the toucan, the sea monster, the arrow, and the flying fish).

Virtually every detail in manuscript and printed constellations reveals some important information about the intentions of the artist. Looking at stars as groupings is rather like looking at clouds and trying to identify the formations with familiar objects; very few clusters of stars actually trace a complete outline of some recognizable person, animal, or object. As a result, celestial cartographers had extraordinary leeway in how they portrayed the constellations. Some constellations have appeared in classical Roman or Greek dress, for instance; some are heavy with soldiers and war figures, reflecting turbulent eras; others glorified the scientific research of the day.

M.J. Fortin. *Atlas céleste* (Paris: F.D. Deschamps, 1776). The 1776 French version of English Astronomer Royal John Flamsteed's celestial atlas of 1725 was the standard star atlas of the late eighteenth century. Flamsteed attempted to correct the many errors he perceived in Bayer's 1603 work, *Uranometria.* Bayer had reversed many of the constellations, showing them from the rear instead of the front, and these new positions contradicted the traditional star descriptions. All of Flamsteed's figures follow the traditional descriptions exactly. (Huntington Library: RB 423014)

The *Poeticon Astronomicon* was one of the primary ancient literary sources on the constellations. During the Renaissance this work was attributed to the Roman historian C. Julius Hyginus, who lived in the first century B.C.E., but is now ascribed to a different Hyginus. The order of the constellations follows the listing in Ptolemy's *Almagest*, so the work may date from the second century C.E. or later. The *Poeticon* was first published in 1475, and the second edition appeared in 1482; the Huntington owns both. The printer Erhard Ratdolt (1447?–1527/8), who had recently moved from Augsburg to Venice, commissioned an elegant set of woodcuts of the constellations for the second edition, and it is the first printed set of illustrated constellations. The figures are somewhat crude but very lively, and viewers have always found them quite charming. Although star positions are indicated, they have little to do with either the positions described by Hyginus, or the actual positions of stars in the sky. Thus it is really not accurate to call these star maps, or the Ratdolt publication a star atlas. But when true star maps did begin to appear in print in the sixteenth century, the set of Ratdolt illustrations provided the initial prototypes for the constellation figures.

Some astronomers took it upon themselves to embed their religious beliefs into widely published star charts. The work of Julius Schiller (d. 1627), a devout German Catholic, reflects an extraordinary marriage of imagination and faith in his interpretation of the heavens. His 1627 work *Coelum Stellatum Christianum* (Christian celestial heaven) was the result of his attempts to completely rename and reconfigure known constellations of the time, converting the skies into a represen-

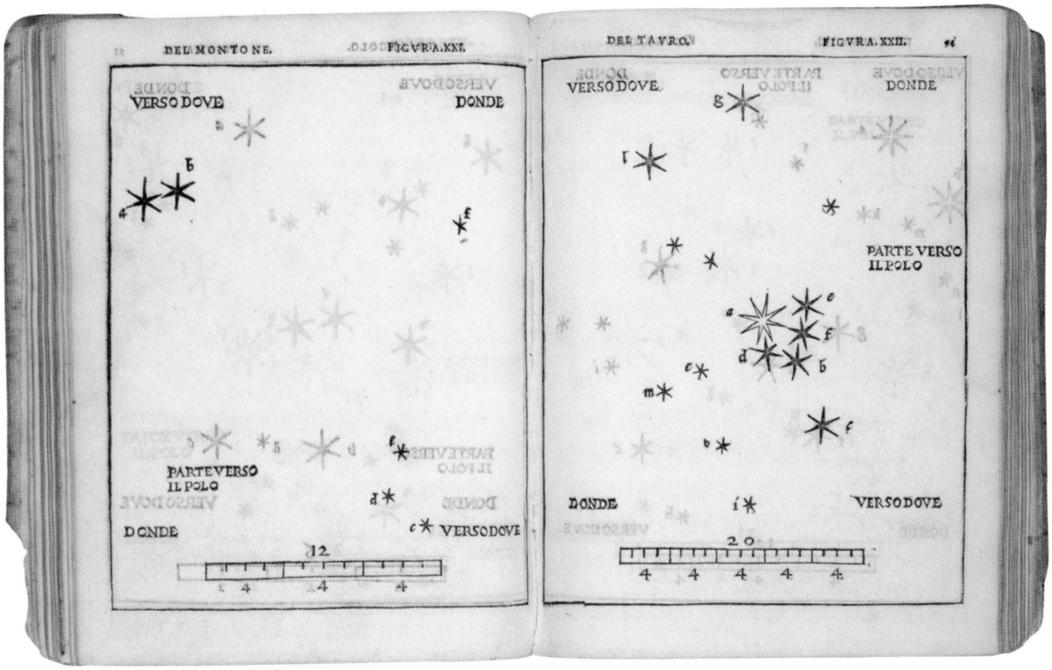

Alessandro Piccolomini. *de le stelle fisse* (Venice: Giouanantonio & Domenico fratelli..., 1540). Piccolomini's volume of constellations contains a total of forty-seven charts (there are forty-eight Ptolemaic constellations, but one was accidentally omitted). Piccolomini made use of a magnitude scale, with brighter stars represented by larger symbols. The Piccolomini work—too small to properly be an atlas, as it was only a large octavo in size—went through at least ten editions between 1540 and 1570. (Huntington Library: RB 482266)

tation of Biblical iconography. His work was produced with painstaking care and as he called upon Tycho Brahe, Johannes Bayer, and Johannes Kepler, among others, for information, it represented the very latest knowledge of the skies.

An Italian humanist and later in life, an archbishop, Alessandro Piccolomini (1508–1578) spent his youth studying and translating literature, and was a great and effective popularizer of science. He wrote several comedies and poems before turning his attention to astronomy. While he wrote other tracts, including philosophical texts, he is best remembered for his work with the stars. A pioneer in the field of astronomy, Piccolomini published what is considered the first printed star chart in 1540, preceding the work of Copernicus, which was published years later. In his book *de le stelle fisse* (On the fixed stars) he presents the first numbering system, using both numerals and Greek letters to denote the relative brightnesses of stars.

Johann Bayer's 1603 *Uranometria* was the first star atlas to represent the stars of the southern latitudes. It is also the first star atlas based on observations made on an actual voyage of discovery—undertaken by Frederick de Houtman's first voyage from the Netherlands to the East Indies in 1595–97. Funded by wealthy Amsterdam investors, the voyage was a disaster; three-quarters of the crew was lost to disease and scurvy. But Houtman's observations of the stars, and Bayer's subsequent illustrations of them, would prove invaluable for future explorers. The work contains forty-nine charts of constellations, many expressed as familiar mythological figures, and two charts of the northern and southern hemispheres. The work included twelve new southern asterisms—informal yet distinctive groupings of stars.

Art historians acclaim Bayer's work for the beauty of figure design and the quality of the engraving. Historians of science are still bringing to light its influence on later developments in celestial cartography. Publication of the beautiful and influential *Uranometria* shaped the way the heavens would be perceived for more than two centuries. Although Piccolomini started the lettering of stars, it was in following Bayer that the practice became widespread and eventually was used by all later astronomers.

Only sixteen copies of John Bevis's rare 1786 *Atlas celeste* are known to exist. The Huntington's copy lacks one of the fifty-two original star charts, but contains the index, which is often lacking in copies. The atlas has an unusual history. John Bevis (1693–1771), an eighteenth-century physician turned astronomer who discovered the Crab Nebula, compiled this atlas between 1745 and 1750, based on Bevis's

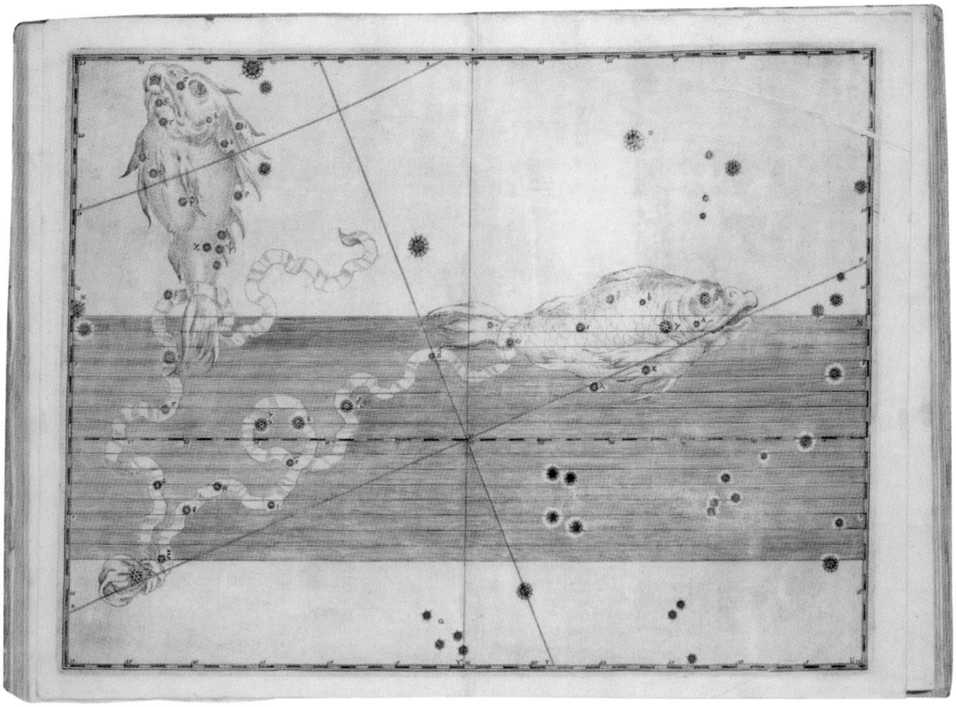

Johann Bayer. *Uranometria: omnium asterismorum continens schemata, nova methodo delineata, aereis laminis expressa...* (Augsberg: Excudit Christophorus Mangnus ..., 1603). Considered the key forerunner of all later star atlases, Bayer's work contains fifty-one star charts, one for each of the traditional forty eight Ptolemaic constellations, along with a chart of the southern skies, based on Dutch explorer Frederick de Houtman's observations,and two planispheres. The maps are steel engravings, rather than woodcuts. Bayer's work is strikingly original; many of his illustrations have no known prototype. (Huntington Library: RB 482143)

own observations in 1738 and the extensive star catalogs of the former Astronomers Royal John Flamsteed (1646–1719) and Edmond Halley (of Halley's Comet fame). Bevis planned to publish the work under the title *Uranographia Britannica* in about 1750 and had obtained subscriptions and private donations from many British and European notables to finance the elaborately illustrated plates. Several prepublication sets of Bevis's star charts were printed in 1749, but in early 1750 his intended publisher, John Neale, was declared bankrupt, the plates were sequestered by the London law courts, and the atlas was never published. Bevis died in 1771, but in 1786 the sets of star charts that had been printed in 1749 were bought at the auction of Bevis's library following the death of his executor, James Horsfall. The anonymous buyer bound the sets of charts and offered them for sale as the *Atlas celeste,* as we know it today. It is apparent in comparing the Bevis atlas to Bayer's *Uranometria* that Bevis followed the plan of the Bayer atlas exactly. There are the same number of plates, each of the same size, and each covering the same area of the sky. The constellation figures are also stylistically identical. But the two are not clones: Bevis has more stars, and more accurate positions for those stars. He also took pains to include the many new or variable stars that had been recently discovered, as well as the nebulous objects. There are nine Messier objects (celestial bodies in the form of nebulae and star clusters, named for Charles Messier (1730–1817), a French astronomer who identified and catalogued 103 of them in the late eighteenth century) on the Bevis charts, and five of them had never before appeared in a star atlas.

Johann Elert Bode (1747–1826) of Germany also made important contributions to celestial atlases, with his *Uranographia* of 1801. His was the first atlas to portray almost all stars visible to the naked eye (down to the sixth magnitude), along with many other stars as faint as the eighth magnitude and visible only through telescopes. He charted about 17,240 stars in more than one hundred star groups. But Bode's atlas marked the end of an era, for he had reached the practical limits of the printed celestial star chart. The artful illustrations of the constellations became a distraction rather than a central feature. Celestial objects had become so numerous that portraying them on paper with any degree of accuracy created a very cluttered sky, and placing them into constellations became less meaningful. Post-Bode atlases have focused on precision measurement of brightness, position, and other physical properties. *DL*

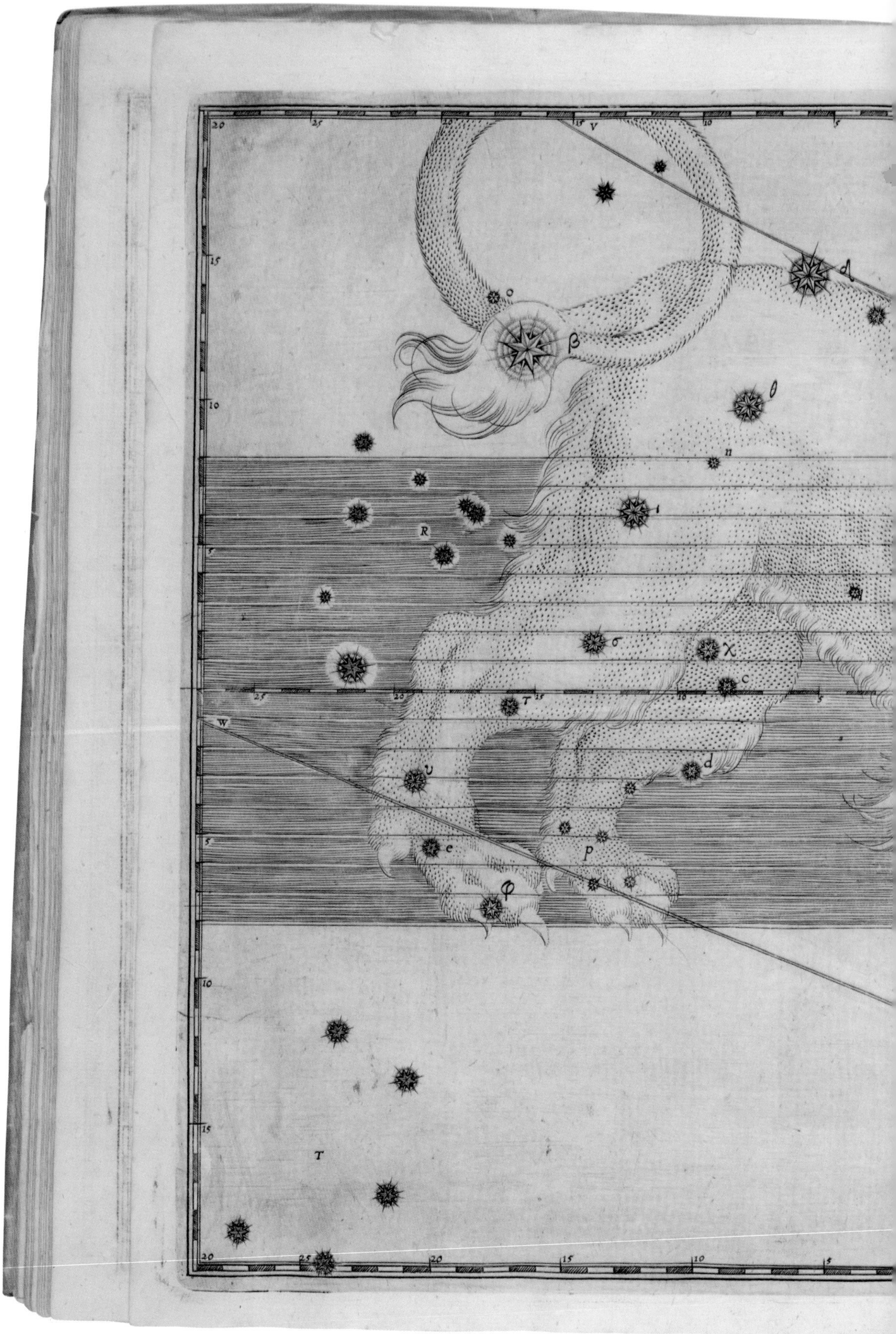

Johann Bayer. *Uranometria: omnium asterismorum continens schemata, nova methodo delineata, aereis laminis expressa...* (Augsberg: Excudit Christophorus Mangnus ..., 1603). Each plate in the Bayer atlas has a grid surrounding the image, allowing star positions to be read off to fractions of a degree. These positions were taken from the catalog of Tycho Brahe, which had circulated in manuscript in the 1590s, but which was not printed until 1602. (Huntington Library: RB 482143)

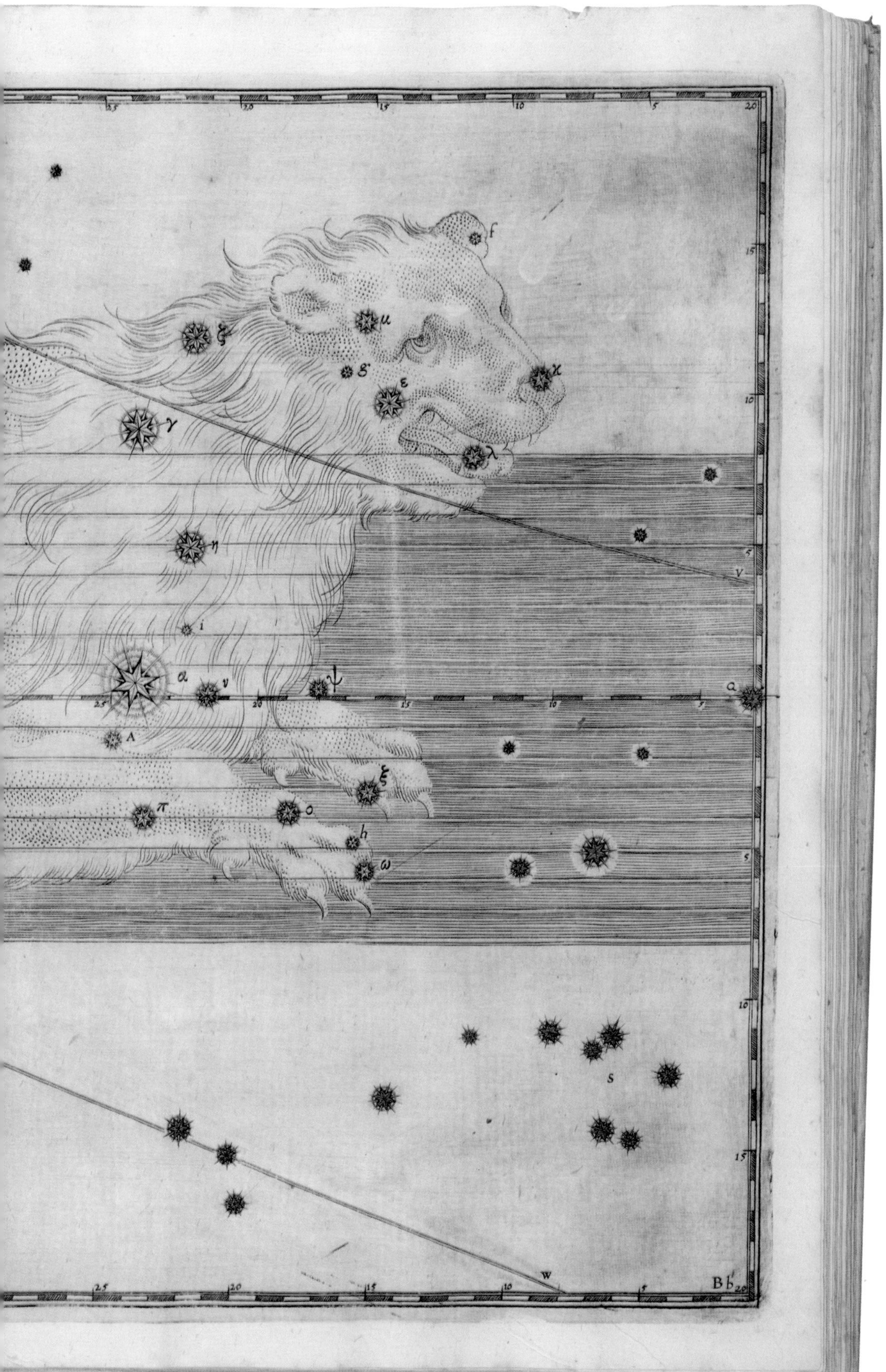

Bb

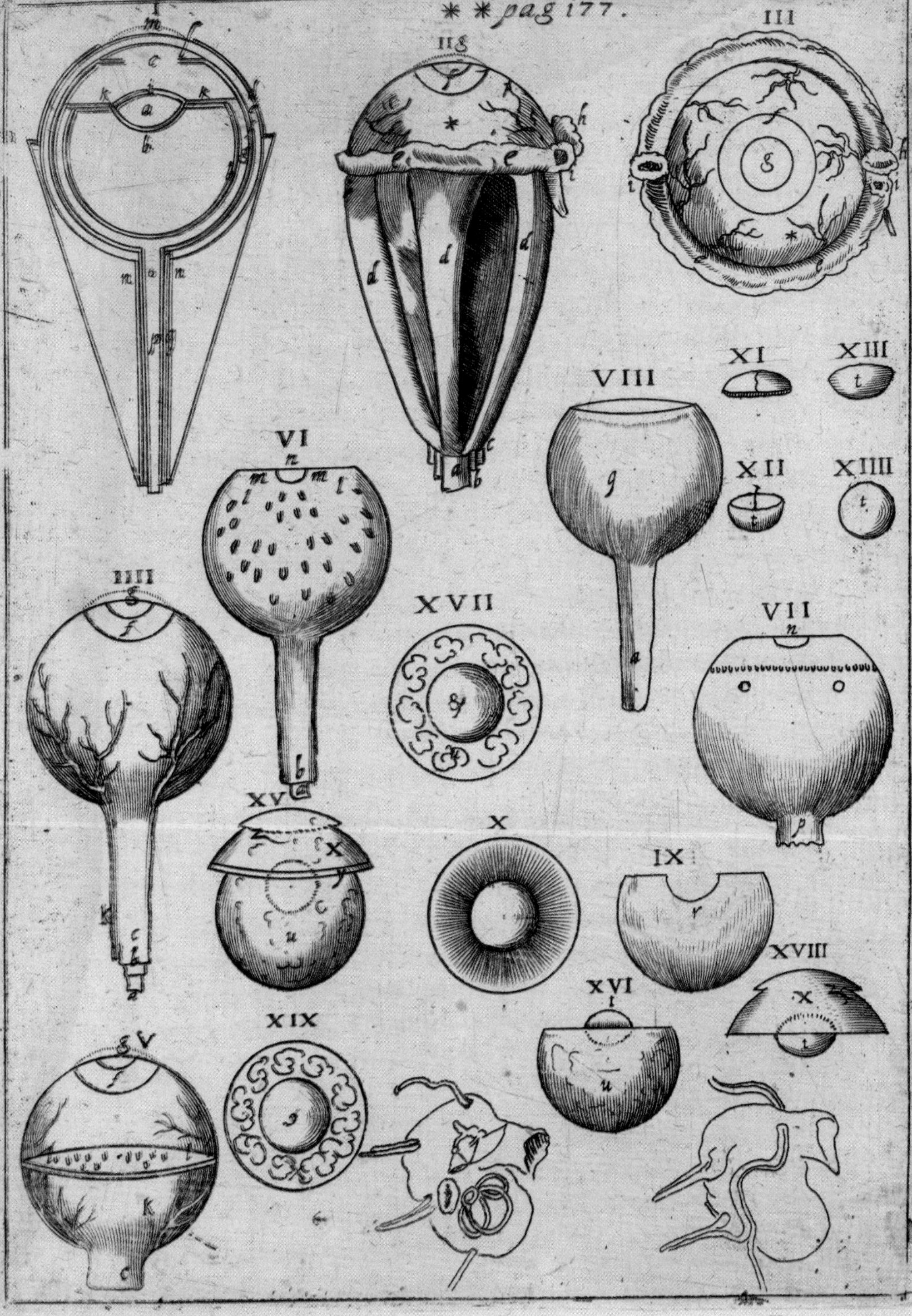
I
II
III
IIII
V
VI
VII
VIII
IX
X
XI
XII
XIII
XIIII
XV
XVI
XVII
XVIII
XIX

TEN | LIGHT AND SPECTRA

WITH THE INTRODUCTION OF THE TELESCOPE in the seventeenth century, astronomers took renewed interest in the nature of light. They would use their knowledge of the behavior of light to produce better instruments and unlock the secrets of the physical nature of the stars. The first telescopes, such as those made and used by Galileo, were just simple spyglasses with lenses producing bright and erect images but with a very narrow field of view. Johannes Kepler was greatly interested in optics, as he was keenly aware of how the Earth's atmosphere bent, or refracted, the light from celestial bodies and caused a change in their apparent position. In his 1604 work, *Astronomiae pars optica* (The optical part of astronomy), Kepler discussed his thoughts on the workings of the human eye and the nature of vision:

> Seeing amounts to feeling and stimulation of the retina, which is painted with the colored rays of the visible world. The picture must then be transmitted to the brain by a mental current, and delivered at the seat of the visual faculty.

Johannes Kepler. *Ad vitellionem paralipomena, quibus astronomiae pars optica traditur...* (Frankfurt: Claudius Marnius & heirs of Johannes Aubrius, 1604). Kepler's study of the eye and its relation to optics forms part of this classic work. He was the first to correctly explain that light entering the eye forms an inverted image on the retina, how eyeglasses cure near- and farsightedness, and why both eyes are needed for depth perception. (Huntington Library: RB 487000.347)

Using his understanding of optics and vision, Kepler described his concept of a refracting telescope which would have a much larger field of view although the image is inverted. The Keplerian telescope is still the basic model used for today's refracting telescopes.

Even though seventeenth-century astronomers had developed a refined notion of refraction, they remained puzzled about the nature of light itself. Some thought it might consist of a stream of particles and others thought light was produced by undulations in the space (aether) that permeated the entire universe. The latter, or wave theory of light, was articulated by Christiaan Huygens, whose influential work *Traité de la lumière* (Treatise on light), was published in 1690. Huygens explained the refraction of light as an effect caused by light slowing down as it entered a more dense media (glass, for example). Huygens derived the laws of reflection and refraction at about the same period as the Dane, Olaus Römer (1644–1710) determined the speed of light itself in 1676 by studying the moons of Jupiter. But even with the creative work done by Huygens, the wave theory of light was not universally accepted until the work of Thomas Young (1773–1829) showed how well the theory explained the behavior of light. Today scientists recognize that light has a dual particle-wave nature.

Isaac Newton's pioneering study of light opened up a new avenue of research for astronomers. In 1665 he looked carefully at the colors produced when light passes through a triangular piece of glass called a prism and concluded that white

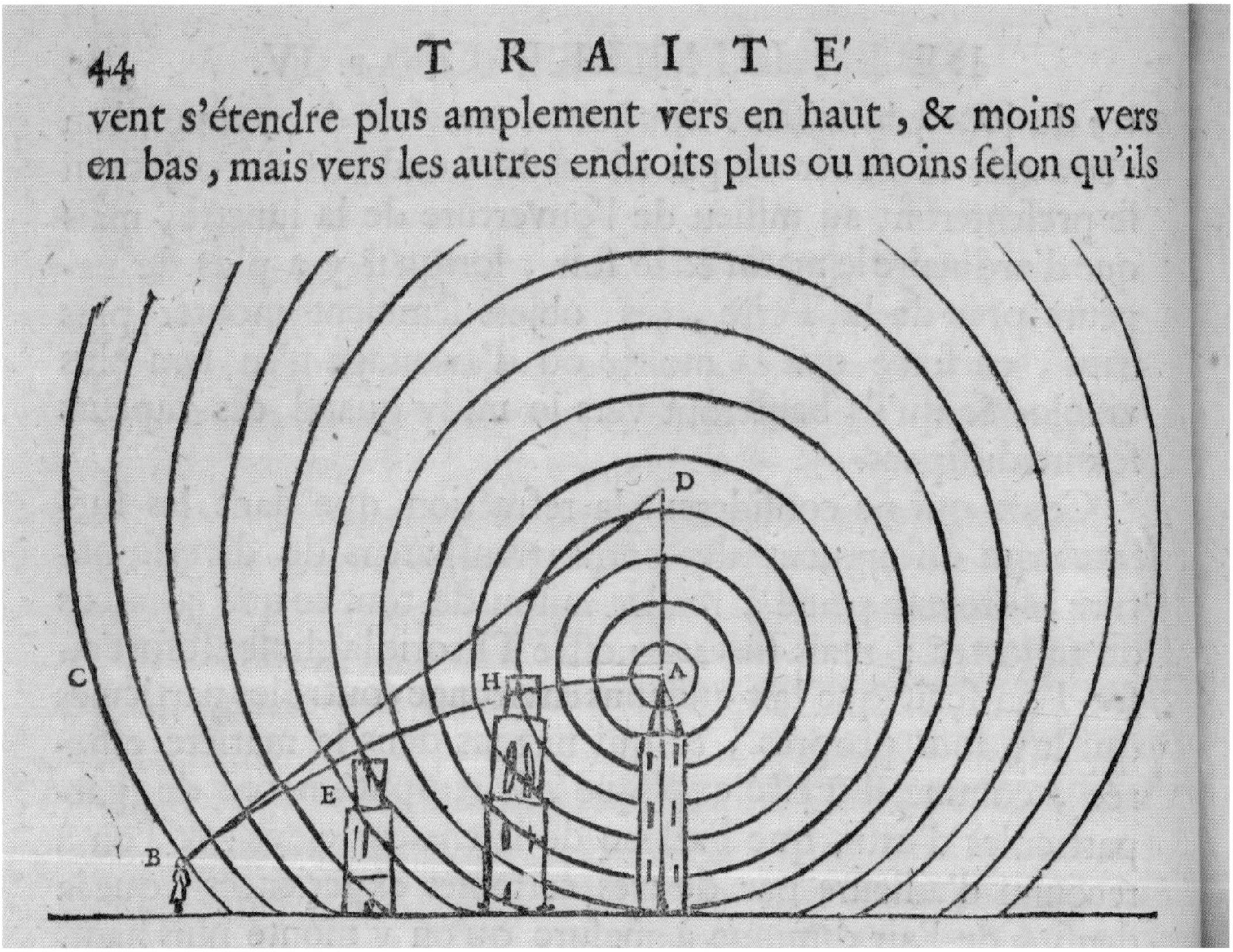
44 TRAITE'
vent s'étendre plus amplement vers en haut, & moins vers
en bas, mais vers les autres endroits plus ou moins ſelon qu'ils
D
A
H
C
E
B

Christiaan Huygens. *Traité de la lumière.* (Leiden: Peter van der Aa, 1690). In this work Huygens formed a beautiful explanation of the reflection and refraction of light in terms of his principle of the motion of wavefronts. This image explains how air can change the direction of light through refraction so that an observer sees a light at point D when it is actually at point A. (Huntington Library: RB 487000.432)

light is made up of a range of colors, which produced a spectrum that was red at one end and violet at the other. The color red was associated with waves of light that were longer than the waves of violet light (hence red light is said to have a longer wavelength than violet light). There were no obvious astronomical applications for this growing comprehension of the structure of white light, in the seventeenth and eighteenth centuries, but the nineteenth century astronomers began to see how light from a celestial body, when broken down into its colored spectrum, would reveal that body's true nature. In the early 1810s Joseph Fraunhofer (1787–1826), while examining the color spectrum produced by sunlight passing through a glass prism, noticed a number of dark lines in the Sun's spectrum. He measured more than four hundred of these lines and noticed similarities between some of these and the bright lines in the spectra produced from the light of glowing chemicals in the laboratory. Around 1860 Gustav Kirchhoff (1824–1887) finally managed to explain that the dark lines in the solar spectrum were all produced by chemical elements and molecules in the Sun's atmosphere. Each element and molecule would absorb one or more very specific colors of light coming from the Sun's interior, which would produce a signature series of dark lines in the spectrum. Identifying them would allow astronomers to determine the chemical composition of the Sun or any star they examined. This knowledge fostered a new branch of astronomy called astrophysics, which studies the physical nature of the stars and planets. Joseph Norman Lockyer (1836–1920) was a pioneering astrophysicist who played an important role as a science popularizer. He produced a large number of works on astrophysics for both general and specialized audiences. Although his own personal theories never became accepted, works such as his *Studies in spectrum analysis* of 1878 were influential for the new generation of astrophysicists.

Johann Christian Doppler (1803–1853) in the 1840s theorized that the exact position of spectral lines would shift slightly to one side or the other depending on whether the Sun or star being observed was moving toward or away from the observer. Doppler, in his 1847 paper, *"Über den Einfluss der Bewegung des Fortpflanzungsmittels auf die Erscheinungen der Aether-, Luft- und Wasserwellen"* (On the influence of motion on the propagation and appearance of aether, air, and water waves), illustrated how a moving source of sound or light would cause the sound or light waves emanating from it to bunch together in the direction of its motion and spread out in its wake (now known as the Doppler effect). Later, Ernst Mach (1838–1916) explained how stars traveling toward us would have their bunched-up spectral lines shifted toward the violet end of the colored spectrum (violet having the shorter wavelengths), while those moving away from us would have their extended lines shifted to the red end (causing the now-famous redshift). The amount of the Doppler line shift would indicate how fast the object was moving along our line of sight. This method of measuring such radial velocities would prove useful, especially in research on the motion of galaxies.

The nineteenth century also saw the discovery of "invisible" light. In 1800, William Herschel (1738–1822) placed thermometer bulbs in the different color regions of the solar spectrum in order to see if there was a variation in the amount of heat produced by a particular color of light. As a control on his experiment, Herschel also placed thermometers beyond the visible ends of the spectrum. In his famous paper, "Experiments on the Refrangibility of the invisible Rays of the Sun," Herschel noted that the thermometer located beyond the red end of the spectrum registered a rise in temperature. Although he did not follow up on this work, he

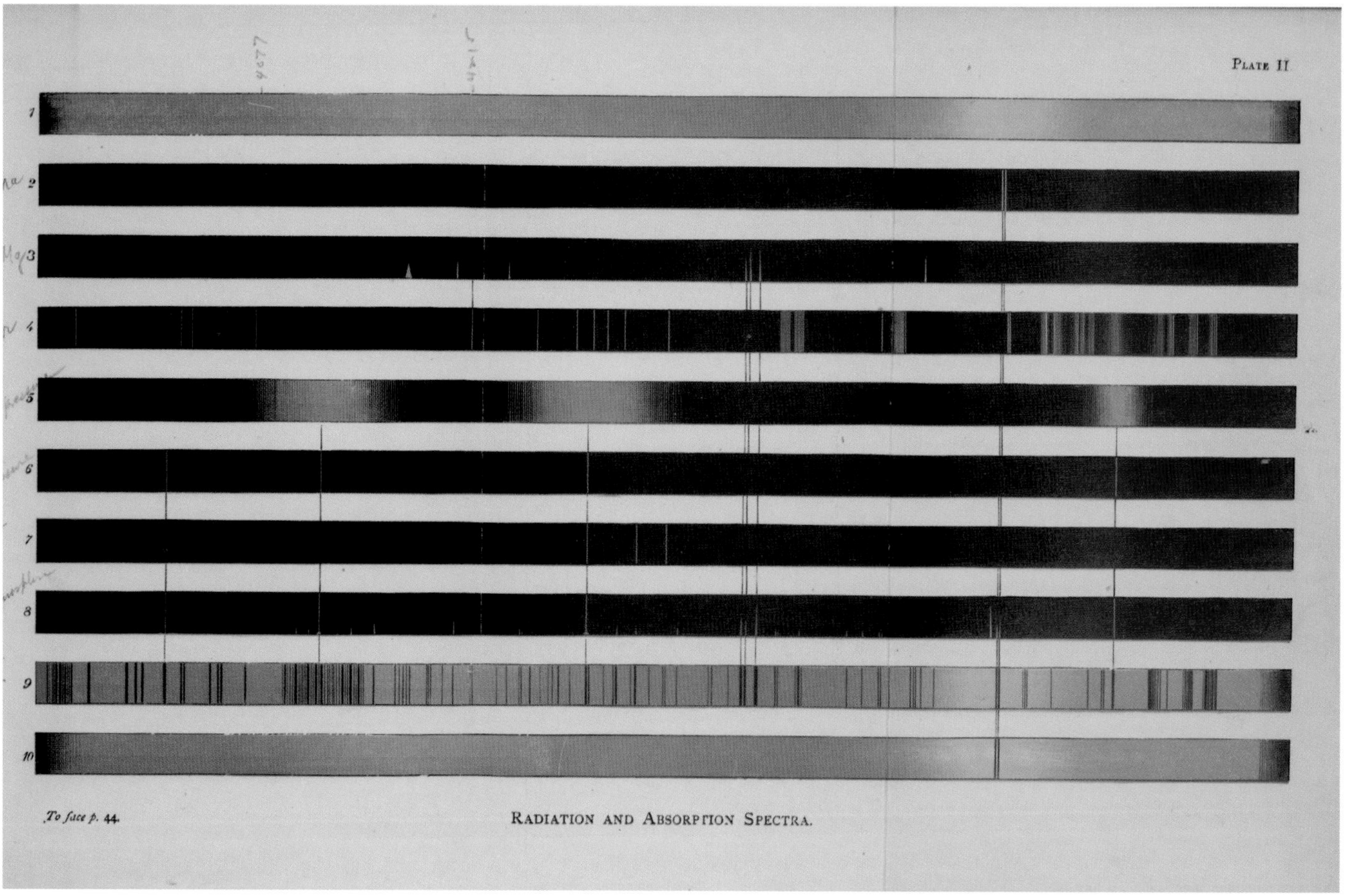

ABOVE

J. Norman Lockyer. *Studies in spectrum analysis* (New York: Appleton, 1878). Here Lockyer shows a number of spectra from various sources, illustrating how spectra can tell us much about the producing object. (Huntington Library: RB 490308)

OPPOSITE

William Herschel. "Experiments on the Refrangibility of the invisible Rays of the Sun," *Philosophical Transactions of the Royal Society*, p. 284. (London: Printed for the Royal Society, 1800). This illustration shows us the layout of Herschel's experiment that discovered infrared radiation. The spectrum produced by the prism at the upper left fell on a bank of thermometers. The unexpected rise of temperature in an invisible part of the spectrum beyond the red light gave away the existence of the infrared rays. (Huntington Library: RB 98681, v. 90)

essentially opened up for examination by astronomers a new, invisible, region of light that came to be known as the infrared region of the spectrum. Scientists also started looking for evidence of light from beyond the violet end of the spectrum and eventually detected "ultraviolet" light in certain laboratory conditions. Eventually visible light was understood to be just a small part of a much larger spectrum of electromagnetic radiation produced by any source of energy. Studying infrared and ultraviolet extensions of the spectra of stars and galaxies would lead astronomers to an even greater understanding of these bodies. The major problem with studying the invisible electromagnetic radiation from beyond the Earth was that our atmosphere tended to absorb most of this radiation before it reached the new electronic radiation detectors that astronomers were developing. Technological achievements in rocketry and the advent of space travel after the World War II allowed astronomers to send their instruments above the intruding atmosphere and finally examine all of the radiation from celestial bodies such as infrared, ultraviolet, gamma, and X-rays. Fortunately, the atmosphere did not obscure astronomical radiation that came in the form of radio waves, and so extremely large dishes, known as radio telescopes, were built that could collect the extremely long-wavelength radio waves from space. This information allowed astronomers to discover such things as the structure of the Milky Way galaxy. *RB*

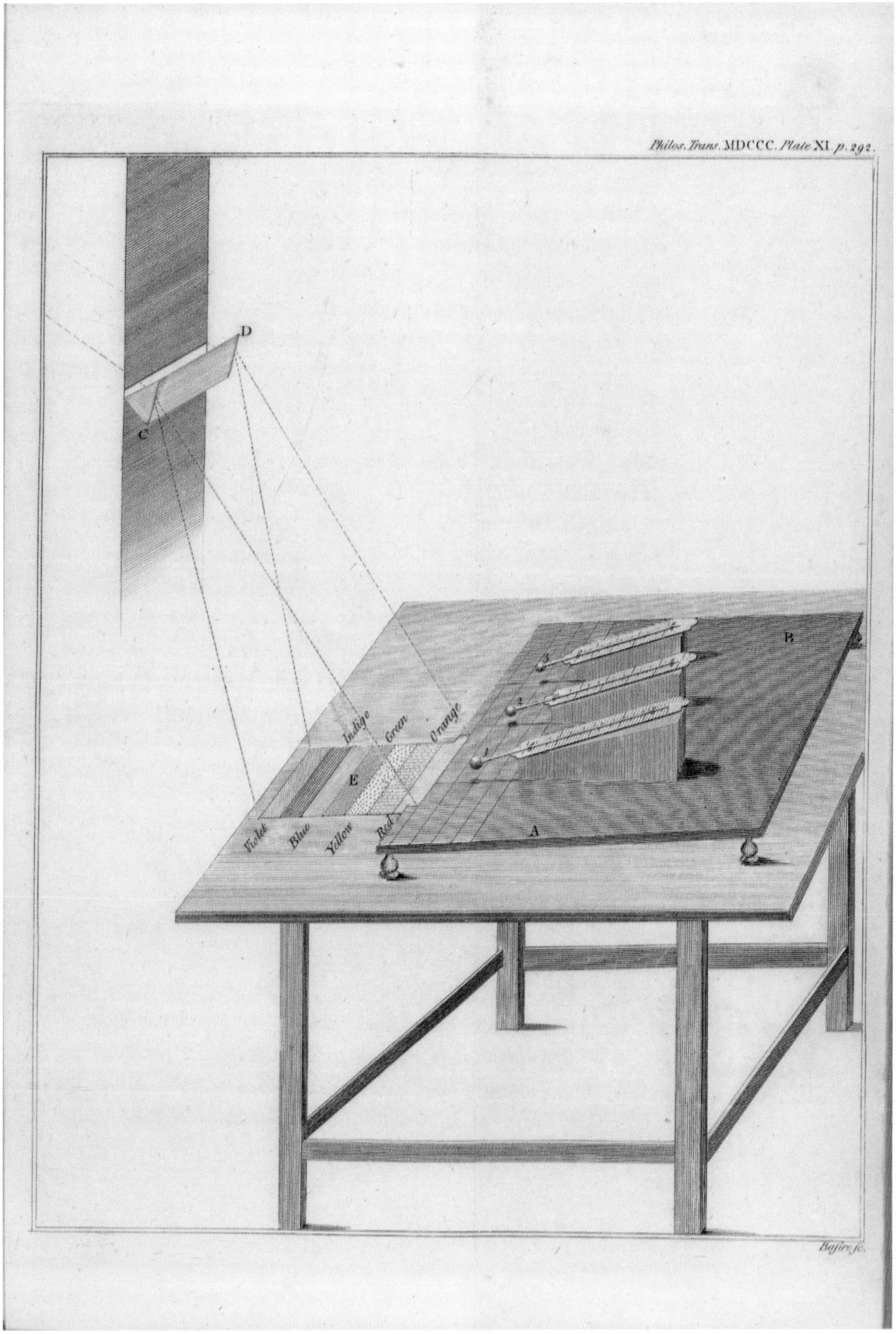
Philos. Trans. MDCCC. Plate XI. p. 292.
D
C
B
A
E
Indigo
Green
Orange
Violet
Blue
Yellow
Red
1
2
3
Basire sc.

SIDEREVS NVNCIVS

MAGNA, LONGEQVE ADMIRABILIA
Spectacula pandens, suspiciendaque proponens
vnicuique, præsertim verò
PHILOSOPHIS, atq; ASTRONOMIS, quæ à

GALILEO GALILEO
PATRITIO FLORENTINO

Patauini Gymnasij Publico Mathematico

PERSPICILLI

Nuper à se reperti beneficio sunt obseruata in LVNÆ FACIE, FIXIS INNVMERIS, LACTEO CIRCVLO, STELLIS NEBVLOSIS,
Apprime verò in

QVATVOR PLANETIS
Circa IOVIS Stellam disparibus interuallis, atque periodis, celeritate mirabili circumuolutis; quos, nemini in hanc vsque diem cognitos, nouissimè Author depræhendit primus; atque

MEDICEA SIDERA
NVNCVPANDOS DECREVIT.

Collegij Soc.is

Regiensis Jesu

VENETIIS, Apud Thomam Baglionum. M DC X.

Superiorum Permissu, & Priuilegio.

1628. di Giorgio Pernicelli

Ex dono D. Bartholomei Pernicelli

ELEVEN | THE EARLY TELESCOPE

THE TELESCOPE HAS TRANSFORMED OUR ABILITIES to observe the heavens more radically than any other single invention in the history of astronomy.[7] From the late thirteenth century until the late seventeenth century, the only tools devised to magnify distances were eyeglasses used to correct poor vision. It is likely that the telescope did exist in some form by the middle of the first decade of the seventeenth century, and perhaps considerably earlier, but historians have had difficulty locating documents that definitively prove its existence, despite tantalizing fragmentary evidence. Three contenders for creating, or at least capitalizing upon the creation of, the first telescope in the late summer and fall of 1608 have been identified. Two were Dutch spectacle-makers: Hans Lipperhey (ca.1570–ca.1619) and Sacharias Janssen (1588–ca.1630); little is known about the third, Jacob Metius except that he and Lipperhey submitted patents to the States-General (the national government in the Hague). All three claimed to be inventors of the telescope, and after study and discussion, the States-General denied the patents to both Lipperhey and Metius, as the telescope was "too easy" to copy" based on the drawings they had presented to the patent office. The telescope was thus never patented and entered the public domain immediately.

Galileo Galilei. *Sidereus Nuncius* (Venice: Apud Thomam Baglionum, March 1610). This landmark work transformed our understanding of astronomy and provided convincing evidence of the "imperfect" nature of the heavens, as illustrated via the telescope for the first time. In the slim volume, Galileo also noted his discovery of four of the moons of Jupiter, which he called the Medicean Stars, dedicating them to the Grand Duke of the Medici family in Italy and the Duke's three brothers. (Huntington Library: RB 487000.71)

Politics and geography of the time combined to make these early efforts widely known. Because the Dutch government was in the midst of negotiating an armistice for its long-running war with the Spanish Crown, The Hague was filled with foreign observers. News of Lipperhey's patent application and descriptions of the telescope itself spread rapidly throughout Europe. Since the invention was simple and easy to duplicate, numerous magnifying devices of the same basic design—primarily spyglasses for use on land, but also primitive telescopes, mounted and meant for stationary use in observing the heavens—were quickly produced. The early models combined convex and concave lenses—as have many telescopes up to the present—but they allowed magnification of only a few power because of the difficulty in accurately pouring and grinding optically accurate lenses.

Although he was neither the actual inventor nor first to use the telescope, the Italian astronomer Galileo Galilei deserves credit for both improving and popularizing it, and was by far its most successful proponent. He created his first telescope in the summer of 1609. He constructed a device of eight or nine power, much stronger than the weak instruments that had appeared previously, which he promptly gave to the chief magistrate of Venice, noting in a letter accompanying the gift that its military applications would be of tremendous value.

Galileo continued to improve the telescope's technology by manufacturing, polishing, and refining his own lenses. His early efforts were impeded by the poor quality of the glass available and the lack of commercial abrasives necessary to grind and polish lenses to satisfactory specifications: only a small percentage of his instruments—approximately 10 percent—actually proved usable. But he continued his efforts. By the fall of 1609, he had turned his attentions to the skies, examining the heavens with a twenty-power telescope and making discoveries that ushered in the age of telescopic astronomy. In December 1609 he made a series of illustrations and sketches of the Moon as viewed through the new instrument, leaving no fewer than eight drawings. The one thing immediately apparent from these illustrations was that under magnification the Moon was clearly not a perfect sphere, but appeared rough and even, quite unlike the "smooth and unchanging" globe that Aristotle had assumed it to be. While Copernican ideas had begun to change notions of an unchanging and perfect universe, few had assimilated all of their implications.

From his correspondence in the fall and winter of 1609, we know Galileo considered his work to be very important and groundbreaking, but within three months, in early 1610, he would make yet another discovery with his telescope that would catapult him to fame: he identified the first four (of what would ultimately prove to be sixteen) moons of Jupiter. Only seven wanderers had been noted amongst the "fixed stars" in the universe since the dawn of time: The Moon and Sun, Mercury, Venus, Mars, Jupiter, and Saturn. Now, suddenly, one of them was shown to have four companions (these four inner moons—Io, Ganymede, Callisto, and Europa—are now known collectively as the Galilean satellites). The discovery also answered one of the key criticisms of Copernican theory: if the Earth wasn't the center of the universe, or at least the solar system, why was it the only visible planet with a moon circling it?

The telescope was increasingly popular, and Galileo knew that others would follow in making improvements and discoveries. In March 1610 he quickly had printed 550 copies of a work called the *Sidereus Nuncius* (Starry messenger), which described his discoveries, including a concise description of how the new instrument functioned, followed by his observations about the Moon and the moons of Jupiter. The work also contained a brief discussion of the fixed stars and describes the difference in appearance between fixed stars and planets. A planned second edition never materialized. The first printing, which sold out in short order, served as the birth announcement for the telescope.

At the same time that Galileo was working to improve the device, Johannes Kepler made the next important advance in the design of the telescope. (The term first appeared in written form in 1611 when Galileo, writing in Latin, coined it the *telescopio*). Kepler discovered that combining two convex lenses would create a much larger field of view and greater magnification—but the image was upside down. The combination of lenses also created more geometric distortion and deviations in color. As lens grinding and production became more sophisticated, these spherical and chromatic aberrations lessened, but the image was still inverted. It was soon discovered that by reflecting the image off a secondary concave lens, the object in the viewfinder could be turned right-side-up again. Much of the second half of the seventeenth century was occupied with attempts to design and build such reflecting telescopes.

In the 1660s Isaac Newton turned his energies toward constructing reflecting telescopes. In 1668 he devised a reflecting telescope that concentrated light by

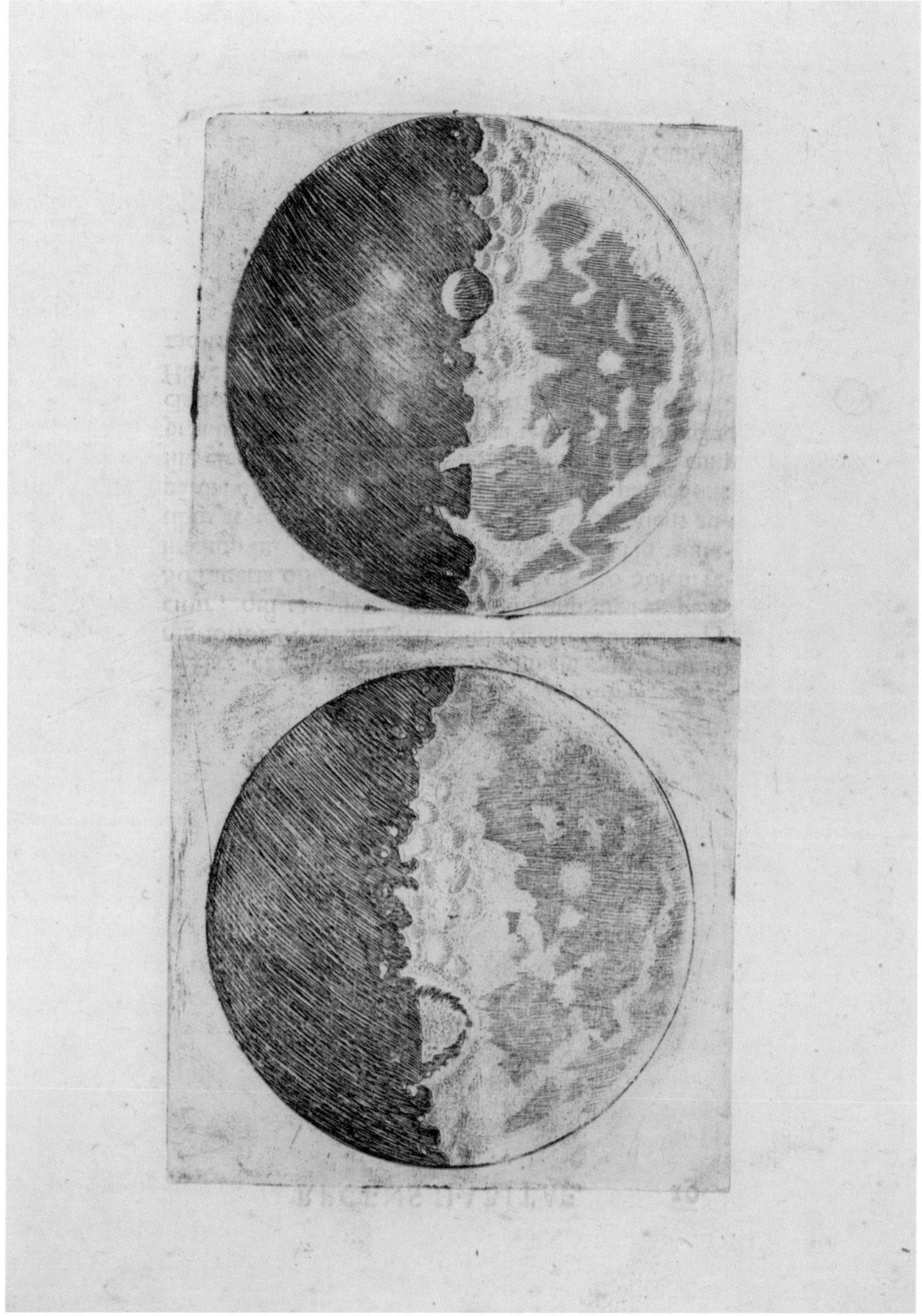

Galileo Galilei. *Sidereus Nuncius* (Venice: Apud Thomam Baglionum, March 1610). The waning gibbous phase and last quarter phase of the Moon. What intrigued Galileo about the Moon was its irregularity, as shown by the new instrument. The large spots on the lunar surface had previously been explained away in various ways: one theory had that the Moon's perfectly smooth parts absorbed and then emitted light in a fashion different from its other parts. (Huntington Library: RB 487000.71)

reflection from a parabolic mirror, rather than by refraction through a lens. This reflecting telescope had another important advantage over refracting telescopes. In Newton's device, light did not actually pass through the surface of the glass, but was reflected off its surface, which allowed for brighter images. Although he produced only small versions with design flaws that would have become evident had he scaled up his models to larger sizes, he became dogmatically convinced that reflecting telescopes were superior to refracting telescopes. The reflecting telescope received the most attention until the 1730s, when improvements in optics led to creation of achromatic refracting telescopes—devices that lessened the distortions in color inherent in the earlier models. *DL*

TO GEORGE THE THIRD KING OF GREAT BRITAIN &c.

This View of a Forty Feet Telescope, constructed under his Royal Patronage, is with permission, most humbly inscribed, by his Majesty's very devoted and Loyal Subject, and most grateful obedient Servant, William Herschel.

TWELVE

MODERN MONSTERS: BIG TELESCOPES

NOT LONG AFTER ASTRONOMERS began using telescopes, they started working on their improvement. The lenses used to collect light were inherently unable to produce perfectly clear images due to distortions in the shape and color of the images (known as spherical and chromatic aberration). These effects could be diminished by reducing the curvature of the lenses (making their surfaces flatter), but as a result the distance from the main lens at which the image came into focus also increased. As they made the lenses flatter, astronomers not only had to stand further and further away from the main lens to observe the image but the images also became dimmer as the focal length of the telescopes increased. Lenses were made larger to collect more light, which in turn required a longer focal length! The Danzig astronomer Johannes Hevelius (1611–1687) eventually constructed telescopes in which the main lens was separated by up to 150 feet from the eyepiece. Seventeenth-century astronomers lacked the technology to mount such a massive tube: Hevelius just made a long wooden plank with the lens at one end and the eyepiece at the other. A tall mast held the telescope, which could be moved around using a system of ropes and pulleys. Unfortunately, the bending of the long plank and vibrations from the wind made such instruments practically useless. Christiaan Huygens and his brother Constantine made even longer telescopes, with focal lengths up to 210 feet. Instead of mounting the primary lens on a plank, they placed it in a small tube at the top of a tall mast. The eyepiece was connected to the tube with a long string, and when the string was pulled taut, the main lens and eyepiece would be aligned. Though this structure was slightly more stable than Hevelius's design, the cumbersome nature of these very-long-focus telescopes rendered them quite inadequate for serious research. An excellent example of such an instrument can be seen in Francesco Bianchini's *Hesperi et phosphori nova phaenomena* (The new phenomenon Hesperus and Phosphorus) of 1728. Bianchini (1662–1729), Pope Benedict XIV's domestic prelate, was a Roman astronomer who used instruments purchased by the Pope from the estate of the famous telescope maker Giuseppe Campani.

Improved reflecting telescopes offered an alternative to the ungainly very-long-focus refracting telescopes (refracting telescopes use lenses to collect light). Isaac Newton's plan of using a mirror to collect the light for use in telescopes was not a practical design for a large telescope, but in the 1720s John Hadley (1682–1744) improved the mirror shape as well as the mounting for a larger reflecting telescope. Further improvements followed, making the shorter reflecting telescope a fine alternative to the long refracting telescopes that used lenses. William Herschel (with the assistance of his sister Caroline, also an accomplished astronomer) pro-

OPPOSITE
William Herschel. "Description of a Forty-Feet Reflecting Telescope," *Philosophical Transactions of the Royal Society*, p. 347. (London: Printed for the Royal Society, 1795). Illustration of Herschel's largest telescope, the 40-feet reflector. The telescope's large tube was supported by a massive framework. A platform at the upper, open end of the telescope allowed the observer to peer into the tube and the mirror at the far end would focus the starlight back into the observer's eyes. (Huntington Library: RB 98681, v. 85).

Joe Hickox at the 60-inch telescope at Mount Wilson Observatory, ca. 1940. (Huntington Library: Carnegie Institution of Washington Collection, COPC 367)

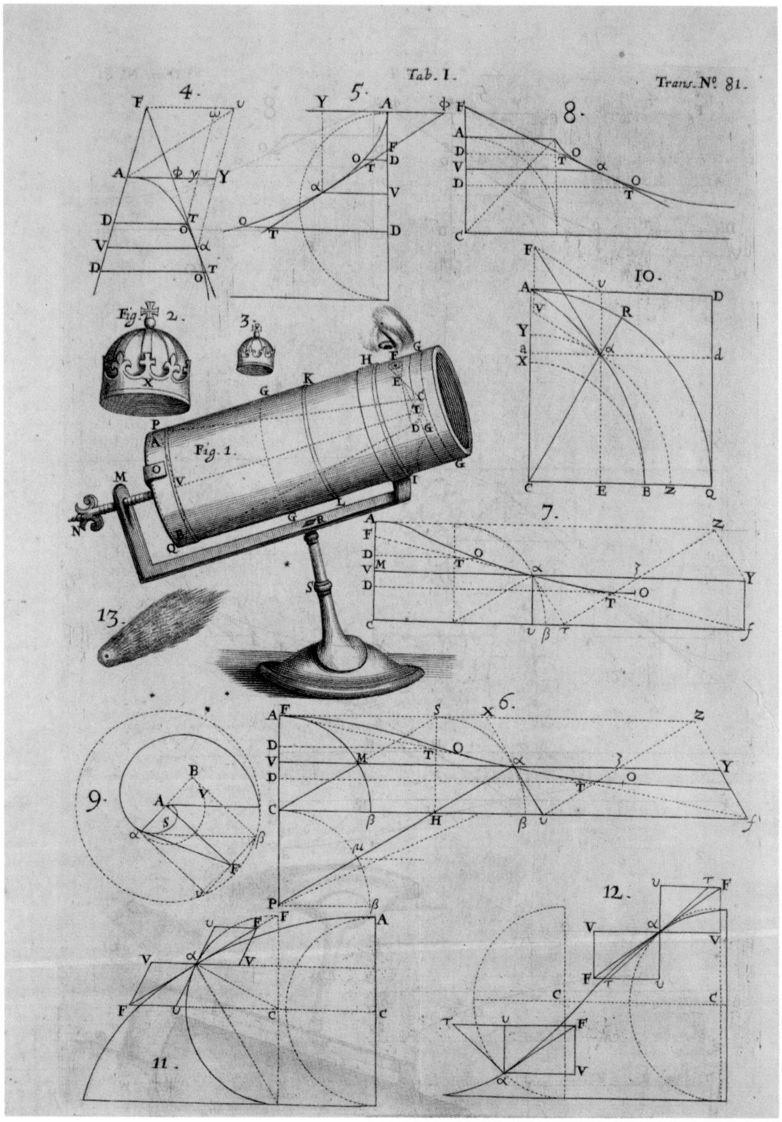

ABOVE
Isaac Newton. "An Accompt of a New Catadioptrical Telescope Invented by Mr. Newton," *Philosophical Transactions of the Royal Society,* p. 4004. (London: Printed for the Royal Society). Catadioptrics involved passing light through a concave lens, then bouncing it off a flat speculum, or mirror, and thence to the eyepiece of the telescope. This eyepiece design, positioned at a 45-degree angle from the mirror, is known as the Newtonian focus and is still in use today. (Huntington Library: RB 98681, v. 7)

OPPOSITE
Giovanni Jacopo Marinoni. *De astronomica specula domestica organico apparatu astronomico...* (1st edition. Vienna: Leopold Johann Kaliwoda, 1745). Observatories often combined living and viewing spaces, with the upper platform providing stable viewing space as well as insulating the room below during different seasons. (Huntington Library: RB 487000.1036)

duced a fine instrument with a mirror 6.25 inches in diameter and a focal length of seven feet, with which he discovered the planet Uranus in 1781. For his success he received a pension from King George III of England that allowed him to continue producing larger telescopes in order to detect fainter and fainter objects. Perhaps his finest instrument was a telescope with an 18.8-inch mirror that was twenty feet in length, but he also built an immense 48-inch mirror in a tube forty feet long, which he described in a paper, "Description of a forty-feet reflecting telescope," of 1795. This instrument never worked as well as Herschel hoped it would and it marked the size limit for telescopes of that period.

Improvements in the quality of glass-making and in optical techniques in the early nineteenth century enabled telescope makers to start producing lenses comparable to the size of the mirror in Herschel's giant reflecting telescope. Alvan Clark (1804–1887) and his sons in Massachusetts created many extraordinary telescopes, including the largest free-moving refracting telescope ever produced, which had a forty-inch diameter lens. George Ellery Hale persuaded a tycoon, C. T. Yerkes, to fund the creation of this telescope for the University of Chicago, and it was installed at Yerkes Observatory in Williams Bay, Wisconsin, which Hale would direct. The lenses for the Yerkes telescope marked the physical limit for refracting telescopes. If the lenses were made much larger they would sag significantly under their own weight and change shape, which would affect their ability to form high-quality images. Reflecting telescopes also improved during this period, through the development of more stable mountings and the switch from metal mirrors to lighter silvered-glass mirrors.

Astronomical research changed radically in the nineteenth century when the new technique of photography was adopted by astronomers. Photography would eventually supplant the human eye as the primary image detector, and the use of long-exposure photography greatly enhanced astronomers' ability to detect even fainter objects. The first significant results were produced by the wealthy British amateur astronomer, Warren De la Rue (1815–1889), who abandoned the daguerreotype process and used a wet-collodion process that was some ten times more sensitive. With the advent of a more convenient and more sensitive dry-plate photographic process, by 1890 photography had become an established part of astronomical research. Photography remained one of the astronomer's most important tools for nearly one hundred years, until it was supplanted by electronic detectors.

In the twentieth century the reflecting telescope totally eclipsed the refractor as the primary tool for visual astronomical research. George Ellery Hale, the man responsible for the largest refracting telescope in the world, realized that reflectors

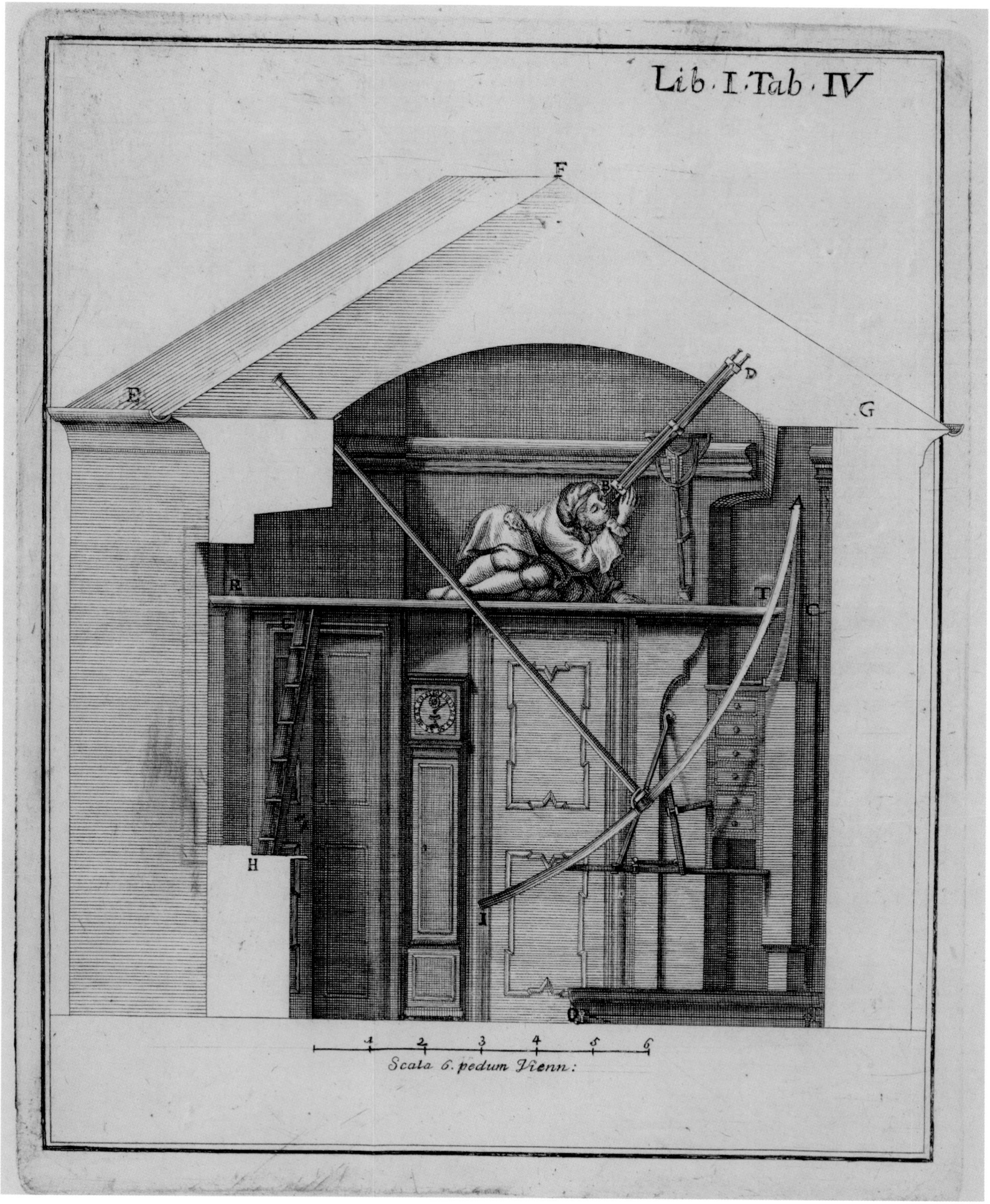
Lib. I. Tab. IV
F
E
G
D
B
R
T
C
H
I
O
1 2 3 4 5 6
Scala 6. pedum Vienn:

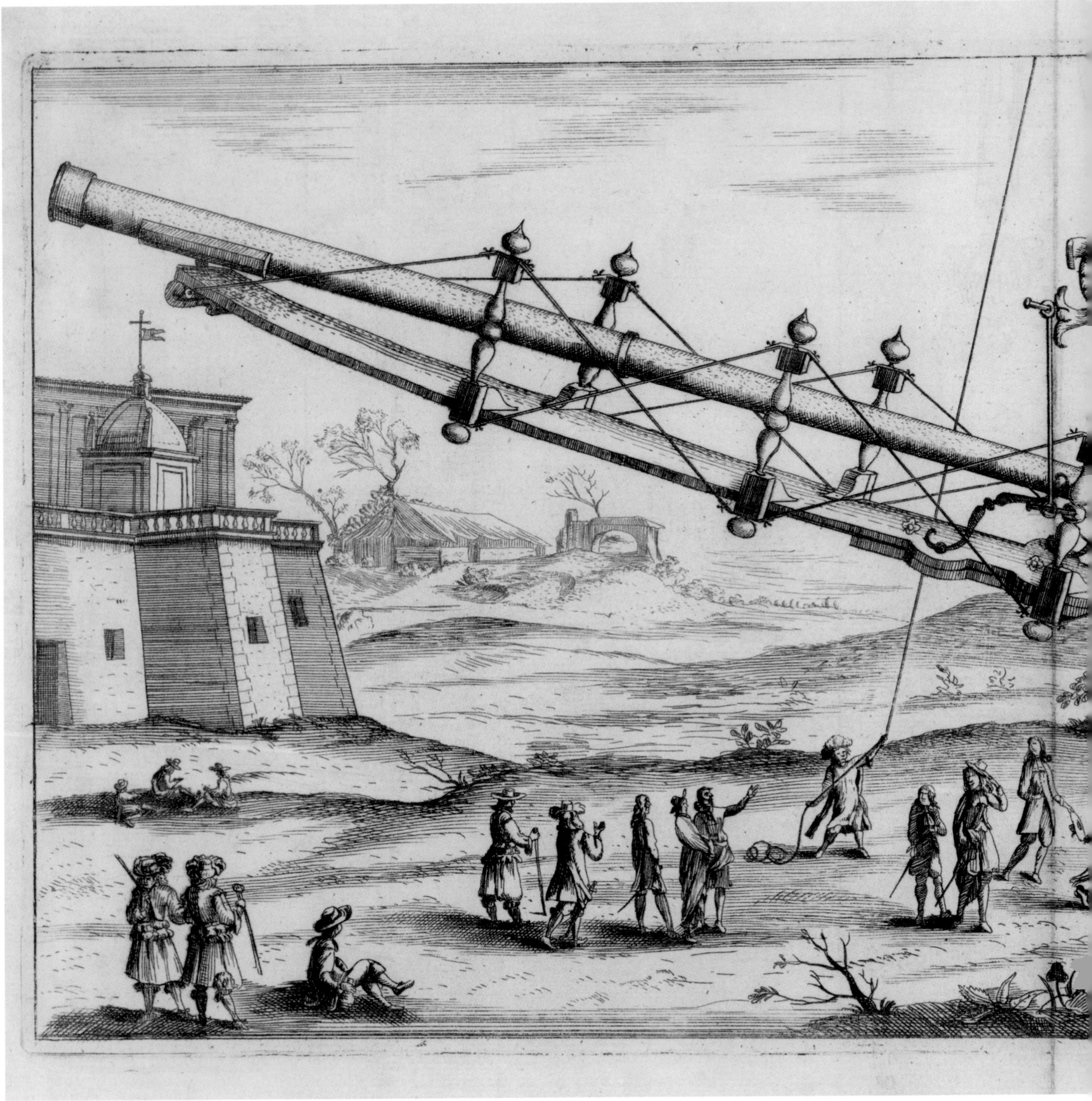

Francesco Bianchini. *Hesperi et phosphori nova phaenomena...* (Rome: Apud Joannem Mariam Salvioni..., 1728). This is an example of the extreme sizes that the long-focus refractors of the early eighteenth century would achieve. The massive size and difficulty in moving the telescope was its primary disadvantage. This particular telescope seems to be suspended from a harness that is not attached to any earthly device! (Huntington Library: RB 475657)

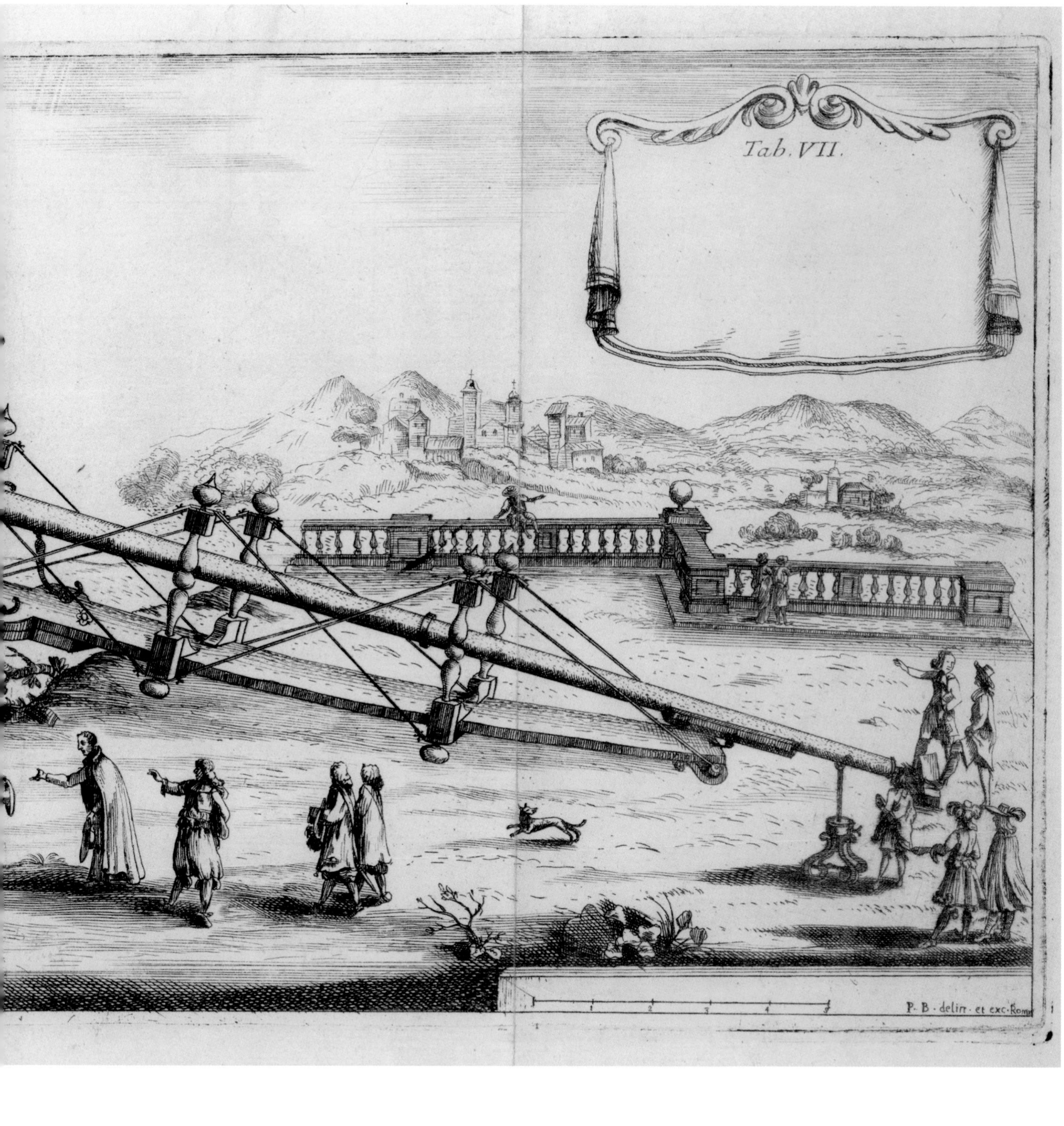
Tab. VII.
1
2
3
4
5
P. B. delin. et exc. Rom

Pierre Borel. *De vero telescopii inventore, cum breve omnium conspiciliorum historia...* (1st edition. The Hague: Adriaan Vlacq, 1655–1656).

TOP: Borel provided an early and excellent detailed image of the Moon, in a work that also contained homage to early telescope inventors Sacharias Janssen and Hans Lipperhey who, he felt, had independently developed the telescope, in that order.

BOTTOM: The sprawling Ursa Major constellation shown here was beautifully interpretated by Borel. The third-largest constellation in the sky, Ursa Major ("Great Bear") is most famous for containing the Big Dipper. All representations of Ursa Major show the bear with a tail—the handle of the Big Dipper—although bears do not actually have tails. (Huntington Library: RB 487000.70)

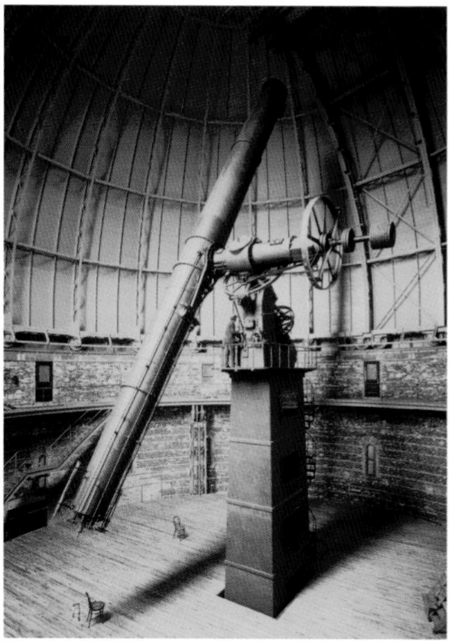

Scrapbook of photographs of Yerkes Observatory, ca. 1897. George Ellery Hale was instrumental in the observatory's creation and served as its director until 1903.

LEFT: Here we see the newly completed 40-inch refracting telescope at Yerkes Observatory, at the time the largest telescope in the world.

The newly completed Yerkes Observatory on the shores of Lake Geneva, Wisconsin in 1897. The 40-inch refractor is housed in the large dome at the far right. Smaller auxiliary telescopes were located in the other domes. (Huntington Library: Hale Papers, uncatalogued)

Mirror blank for the 200-inch Mount Palomar telescope, being delivered by rail from New York to Pasadena, April 1936. After an unsuccessful first attempt, the Corning Glass Works was able to produce a Pyrex glass disk for the 200-inch Palomar telescope, ground into the proper shape at the optical shop at the California Institute of Technology (Huntington Library: Carnegie Institution of Washington Collection, COPC 2445)

had the potential to be made larger and larger and that they marked the path to future research. Accordingly, he secured financing from the Carnegie Institution of Washington for the construction of a reflecting telescope with a five-foot diameter mirror on a massive mounting in a protected dome on top of Mount Wilson, California. The five-foot reflector was first used in 1908 and heralded the coming of the monster telescopes. Hale himself was responsible for raising the money to construct the next two largest telescopes in the world, the 100-inch reflector at Mount Wilson (1917) and the 200-inch reflector on Mount Palomar (1948). For a while, the 200-inch reflector marked the limit of telescope design due to increasing difficulties in mounting such heavy telescopes and in supporting the mirror so that its own weight would not distort its shape. With the advent of computer-correction capabilities and new materials and processing approaches, however, mirrors could be made that were much lighter or even segmented, and larger and larger telescopes again began being built. The 200-inch Palomar telescope is now dwarfed by several instruments all over the world, including the current record-holders of the W. M. Keck Observatory, a pair of telescopes each with segmented mirrors with an equivalent diameter of ten meters. These telescopes, Keck I and II on Mauna Kea, Hawaii, will eventually operate in tandem as if they both were one even larger telescope. Even larger telescopes and telescope array projects are underway, including the four 8.2-meter telescope array of the VLT (Very Large Telescope) operated by the European Southern Observatory in Chile and the near-twin 11-meter wide Hobby-Eberly Telescope in Texas and the South African Large Telescope. Telescopes can now be placed above the blurring and obscuring atmosphere of the Earth: the Hubble Space Telescope, similar in size to the 100-inch reflector on Mount Wilson, currently orbits in space sending back some of the most incredible images of celestial bodies ever seen by humans. *RB*

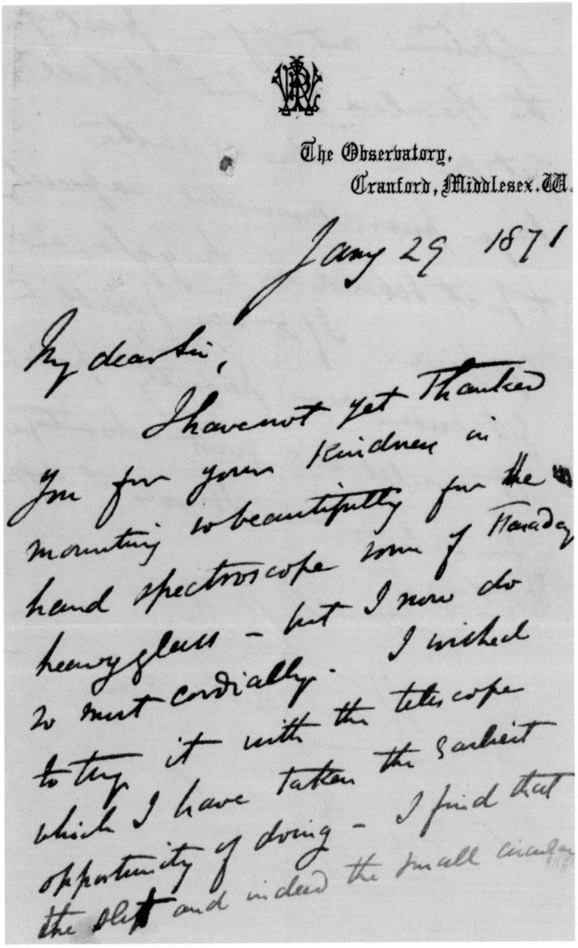

The Observatory,
Cranford, Middlesex. W.

Jany 29 1871

My dear Sir,

I have not yet thanked you for your kindness in mounting so beautifully for me the hand spectroscope some of Faraday heavy glass – but I now do so most cordially. I wished to try it with the telescope which I have taken the earliest opportunity of doing – I find that the slit and indeed the small circular

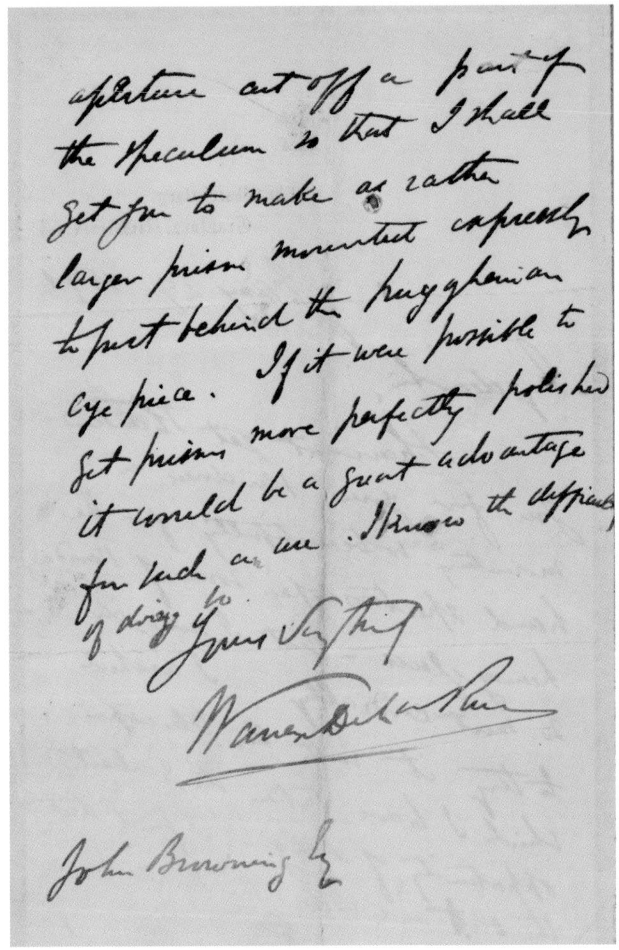

aperture cut off a part of the speculum so that I shall get you to make a rather larger prism mounted expressly to put behind the Huyghenian eye piece. If it were possible to get prisms more perfectly polished it would be a great advantage. I know the difficulty of doing so for such a use.

Yours very truly

Warren De la Rue

John Browning Esq

ABOVE LEFT AND RIGHT

Warren De la Rue to John Browning, 29 January 1871. (Cranford, The Observatory). De la Rue thanks Browning for beautifully mounting a hand spectroscope and notes he hopes to try it with a telescope. (Huntington Library: Uncatalogued)

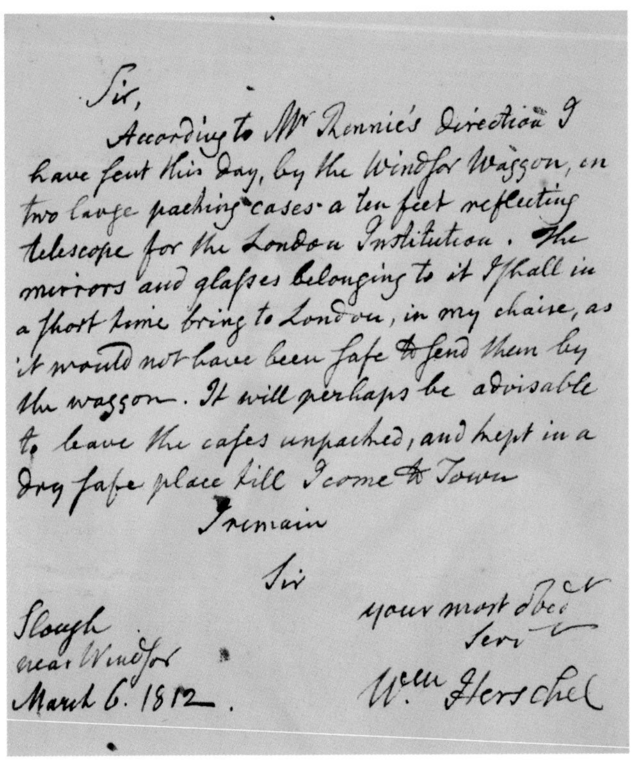

Sir,

According to Mr Rennie's direction I have sent this day, by the Windsor Waggon, in two large packing cases a ten feet reflecting telescope for the London Institution. The mirrors and glasses belonging to it I shall in a short time bring to London, in my chaise, as it would not have been safe to send them by the waggon. It will perhaps be advisable to leave the cases unpacked, and kept in a dry safe place till I come to Town

I remain

Sir

your most obedt Servt

Wm Herschel

Slough near Windsor March 6. 1812.

William Herschel to unknown recipient, 6 March 1812, concerning a 10-foot refracting telescope he is having shipped to London. (Huntington Library: HM 20605)

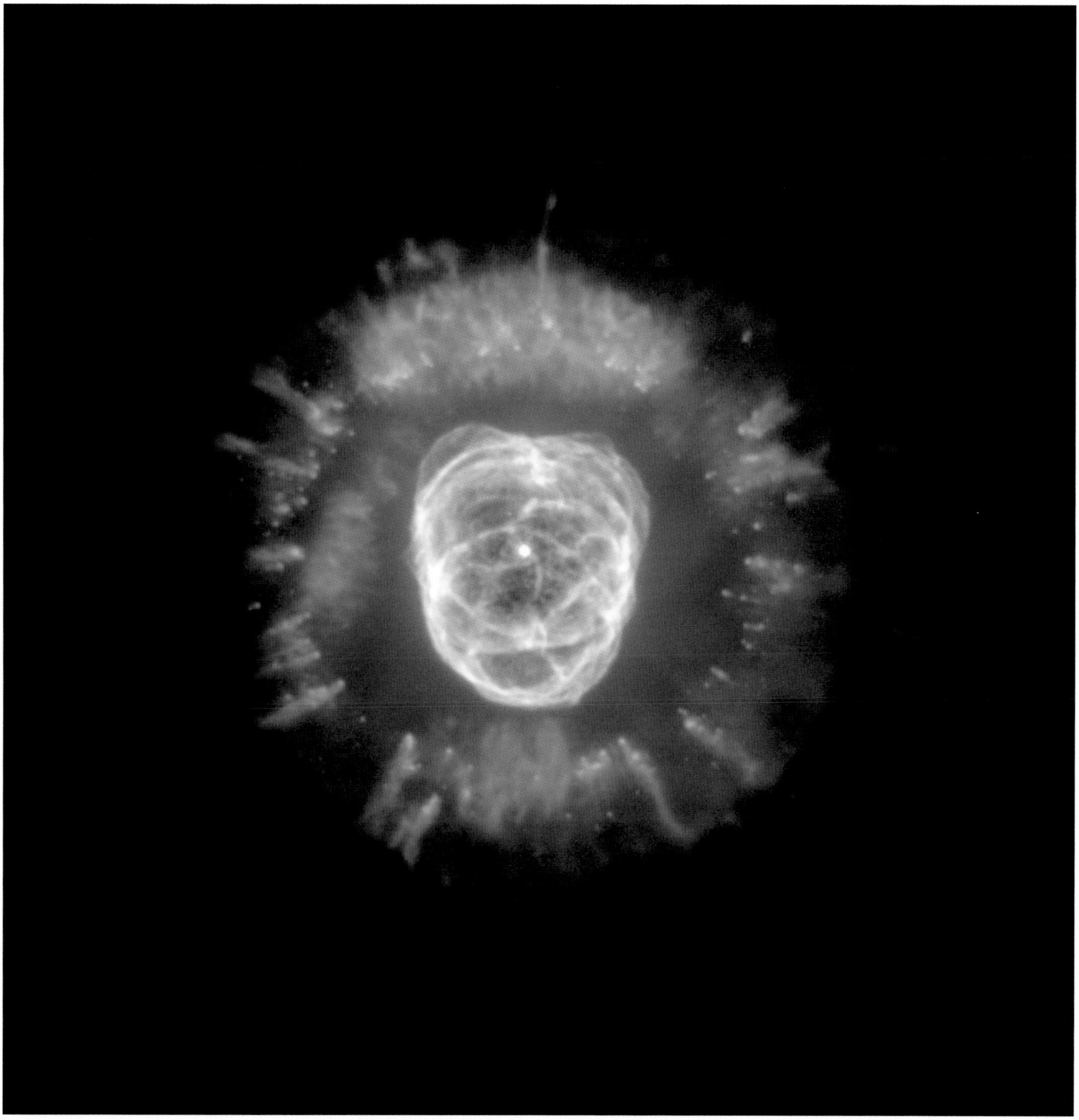

This image of the "Eskimo" Nebula, taken in early 2000 by NASA's Hubble Space Telescope provides a majestic view of a planetary nebula, the glowing remains of a dying, Sun-like star. This stellar relic, first spied by William Herschel in 1787, is nicknamed the "Eskimo" Nebula (NGC 2392) because, when viewed through ground-based telescopes, it resembles a face surrounded by a fur parka. In this image, the "parka" is really a disk of material embellished with a ring of comet-shaped objects, with their tails streaming away from the central, dying star. (NASA, A. Fruchter and the ERO Team [STScI])

III

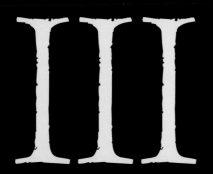

Encountering the Universe

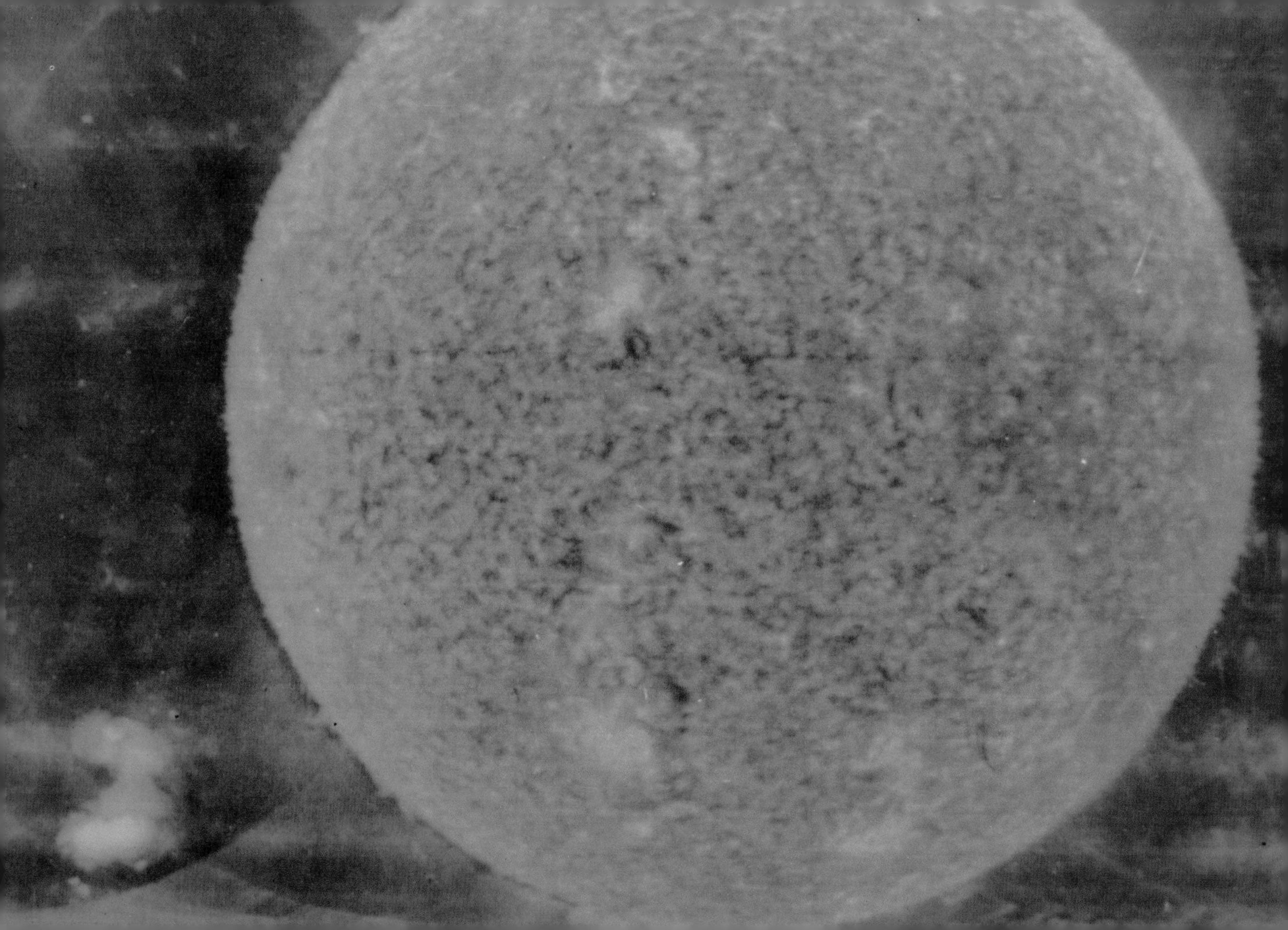

THIRTEEN THE SUN

In June 1973, astronauts on Skylab, America's first experimental space station, photographed the Sun and its prominences—masses of cloudlike gas emitted from a layer of the Sun's atmosphere. (NASA Image)

WHILE THE PLANETS OF THIS SOLAR SYSTEM and celestial bodies in the further reaches of space are generally visible at night, only the Sun is most remarkably visible by day.[8] The Sun is, in fact, what makes itself visible. It is the source of virtually all light and heat on Earth, has been scrutinized from the earliest times, can be felt without even being seen, and continues to provide new discoveries and scientific insights into the twenty-first century.

Modern solar science began in 1610 when Galileo and others aimed their telescopes at the Sun. While solar eclipses had been known for centuries, the Sun itself could not be carefully and systematically viewed without magnification, filters, and other tools. Early views of the Sun through the telescope confirmed the Copernican view that the heavens were changeable: its surface varied with sunspots; it appeared elliptical rather than perfectly spherical; and it exhibited other characteristics that went against Aristotelian and Ptolemaic conceptions of the universe.

The earliest telescopic observations of the Sun—and the source of an important rivalry—were made by two men, Galileo Galilei (1564–1642), an Italian mathematics professor, and Christopher Scheiner (1575–1650), a Jesuit priest and mathematics professor at the University of Ingolstadt. The two men were the principal contenders in an argument over the nature of sunspots—Galileo claiming that the small dark marks were cloudlike formations hovering very close to the Sun, and Scheiner insisting that the spots were actually planets visible against the brightness of the Sun. The controversy produced several important publications, including Galileo's 1613 *Istoria alle macchie solari. . .* (Letters on sunspots), which included his first public endorsement of the Copernican hypothesis, as well as a work by Scheiner called *Sol Elliptic[us]* (Elliptical Sun) published in 1615. The debate was fraught with import for the two men personally, involving a bitter and ultimately empty priority dispute over who actually saw the sunspots first. In actuality, Thomas Harriot (1560–1621) in Oxford, Johann Fabricus (1587–1616) in Wittenberg, and Domenico Passignani in Rome all also observed sunspots at approximately the same time, as telescope technology had spread rapidly.

By the end of the seventeenth century, astronomical telescopes had advanced considerably. Consisting entirely of convex lenses, they had provided a sizeable field of view at magnifications of fifty and above. Telescope manufacturers began adding micrometers for measuring small angles and crosshairs for fine alignments. New discoveries followed as the eighteenth century progressed. The English astronomer John Flamsteed and the Italian astronomer Giovanni Cassini (1625–1712) worked out a fresh means of calculating the Sun's distance from Earth via observa-

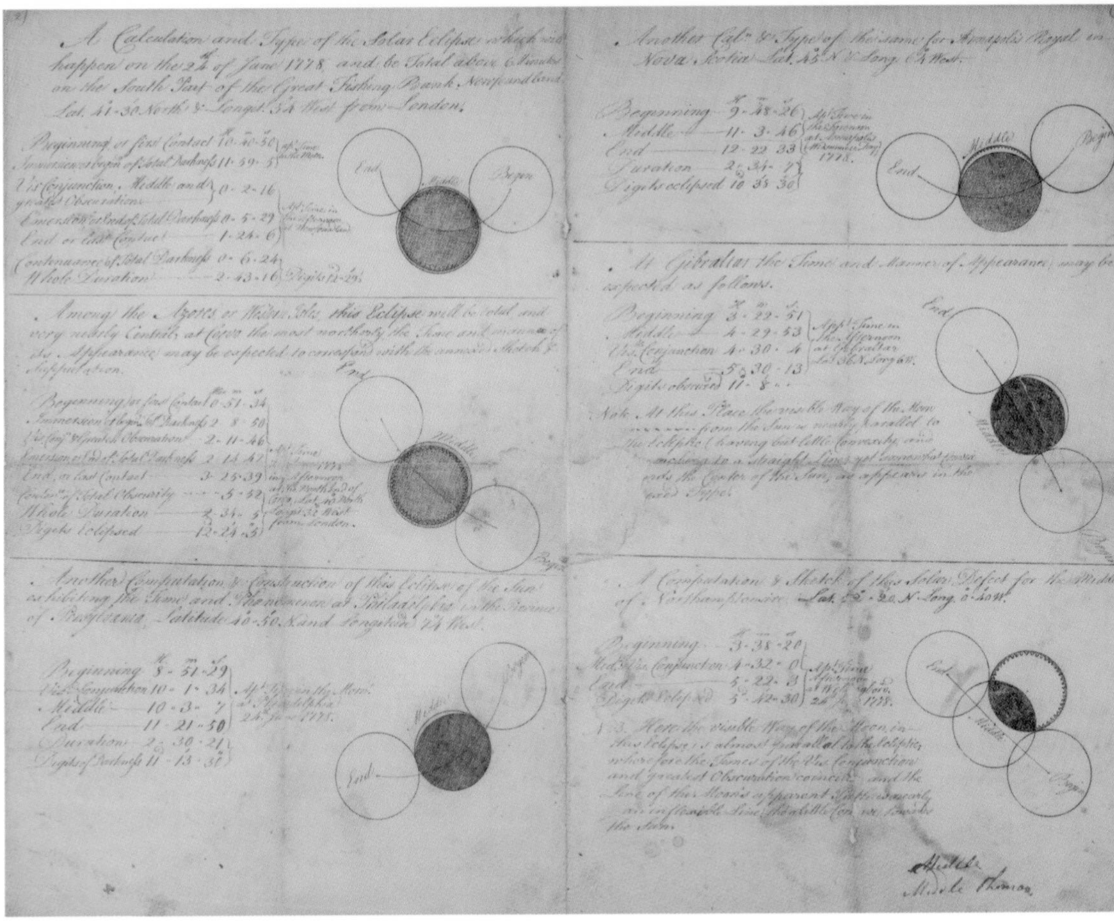

ABOVE AND OPPOSITE
Thomas Cowper. "An Account of Some Astronomical Phaenomena that occur in the years 1776 and 1778." This manuscript shows drawings by Cowper of solar eclipses as observed in England, Nova Scotia, Pennsylvania, and Newfoundland, on 24 June 1778. An extensive network of solar observers around the world had established a long tradition of sharing information about eclipses, and Cowper detailed their appearances and timings, no doubt with help from this network of observers. (Huntington Library: H. H. Sinclair Collection).

tions of Mars. Others attempted to solve problems in solar physics through advances in mathematics, such as the great British astronomer Isaac Newton, who worked on determining the mass and density of the Sun, a process he demonstrated in his *Principia Mathematica* (first published in 1687) and refined in the 1713 and 1726 editions.

The eighteenth century brought new opportunities. Several hundred astronomers from around the world studied the transits of Venus across the Sun in 1761 and 1769, and generated collective measurements that helped to refine estimates of the Sun's distance from Earth, coming within 10 percent of current estimates (92,955,820 miles, a figure now accurate to within a few miles). The great British astronomer William Herschel, best known for his large telescopes and his discovery of the planet Uranus, worked steadily at unraveling some of the mysteries of the Sun. He thought that sunspots were "holes" in the Sun that gave a glimpse of a solid core below a brilliant exterior, that "under the outward flaming Surface is most probably contained a solid globe of unignited Matter."

By the first decade of the nineteenth century, astronomers had a fairly good working knowledge of several important properties of the Sun. They knew its approximate distance from Earth, its basic dimension, mass and density, and even its rate of rotation, which they obtained by tracking sunspots across the solar disk. They were even able to calculate its velocity through space from a slowly increasing body of knowledge about the motions of nearby stars. But new discoveries have a way of meddling with existing conceptions of religion and man's place in

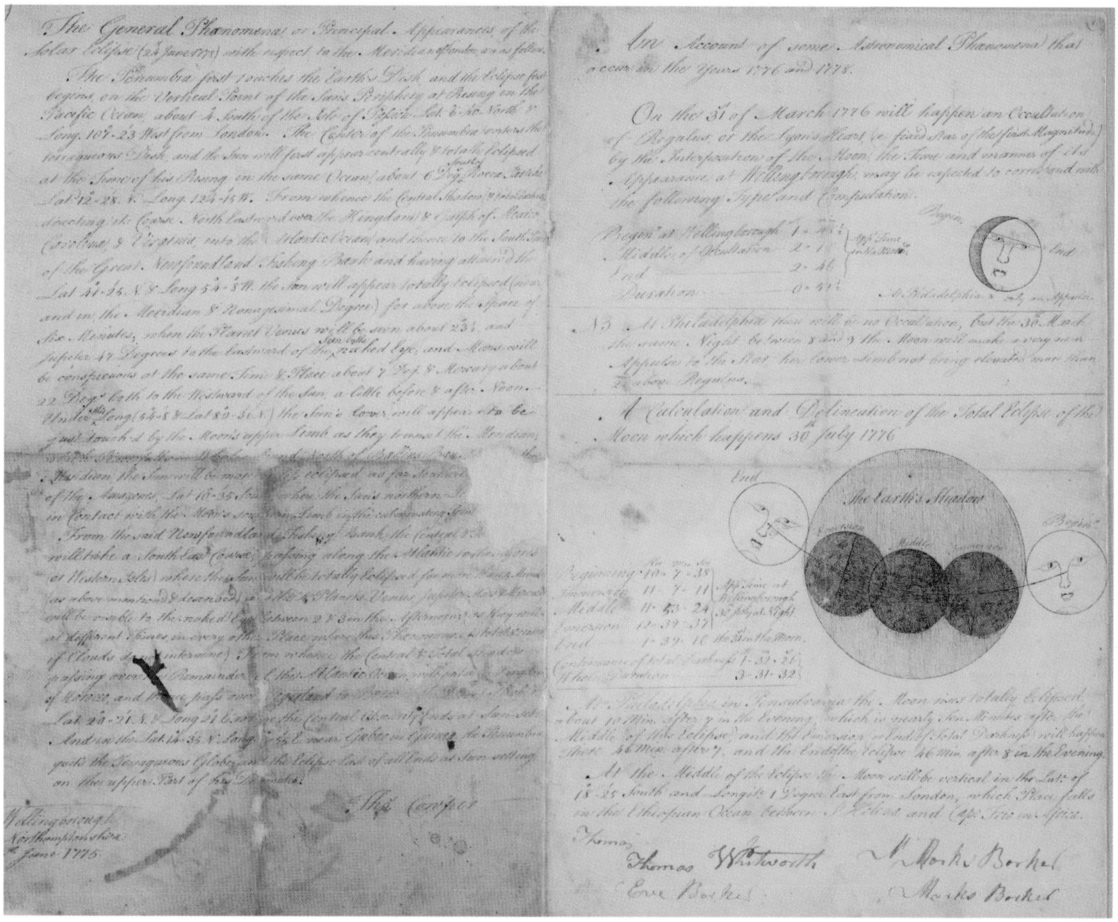

The General Phænomena or Principal Appearances of the Solar Eclipse (24 June 1778) with respect to the Meridian of London, are as follow.

The Penumbra first touches the Earth's Disk and the Eclipse first begins on the Vertical Point of the Sun's Periphery at Rising on the Pacific Ocean about 4 South of the Isle of Pepper Lat. 6-20 North & Long. 107-23 West from London. The Center of the Penumbra enters the terraqueous Disk and the Sun will first appear centrally & totally Eclipsed at the Time of his Rising on the same Ocean about 6 Deg. ... Lat. 12-28 N. Long. 124-15 W. From whence the Central Shadow ... directing its Course North Eastward over the Kingdom & Gulph of Mexico, Carolina & Virginia, into the Atlantic Ocean and thence to the South ... of the Great Newfoundland Fishing Bank and having attain'd the Lat. 41-25 N & Long 54-5 W. the Sun will appear totally Eclipsed ... and in the Meridian & Nonagesimal Degree) for about the Space of Six Minutes, when the Planet Venus will be seen about 23½ and Jupiter 47 Degrees to the Eastward of the Sun by the naked Eye, and Mars will be conspicuous at the same Time & Place about 7 Deg. & Mercury about 22 Deg. both to the Westward of the Sun a little before & after Noon. ...

An Account of some Astronomical Phænomena that occur in the Years 1776 and 1778.

On the 31st of March 1776 will happen an Occultation of Regulus or the Lyon's Heart (a fixed Star of the first Magnitude) by the Interposition of the Moon, the Time and manner of its Appearance at Wellingborough may be expected to correspond with the following Type and Computation.

Began at Wellingborough	1 - 30½
Middle of Occultation	2 - 1
End	2 - 46
Duration	0 - 51½

N.B. At Philadelphia there will be no Occultation, but the 30th March the same Night between 8 and 9 the Moon will make a very near Appulse to the Star ... Regulus.

A Calculation and Delineation of the Total Eclipse of the Moon which happens 30th July 1776

Beginning	10 - 7 - 38
Immersion	11 - 7 - 11
Middle	11 - 53 - 24
Emersion	12 - 39 - 37
End	1 - 39 - 10
Continuance of total Darkness	1 - 32 - 26
Whole Duration	3 - 31 - 32

At Philadelphia in Pensylvania the Moon rises totally Eclipsed about 10 Min. after 7 in the Evening, which is nearly Ten Minutes after the Middle of the Eclipse, and the Emersion or End of Total Darkness will happen there 56 Min. after 7, and the End of the Eclipse 46 min. after 8 in the Evening.

At the Middle of the Eclipse the Moon will be vertical in the Lat. of 18-25 South and Longit. 1 Degree East from London, which Place falls in the Ethiopian Ocean between St. Helena and Cape ... in Africa.

Wellingborough
Northamptonshire
June 1775

Thomas Whitworth

the universe, whatever the era. For instance, the new century brought with it a whole host of new technical observations about the Sun, along with new understandings of the laws of thermodynamics and the idea that things—even the Sun—run down unless they are replenished in some way. Despite vigorous disagreements on the topic (such as the actual age of the Sun and thus its likely lifespan), the notion troubled both evolutionists and creationists, because it undermined the idea of man's place as a survivor based on fitness as well as man's place as a chosen creation of God. Charles Darwin himself noted in 1876: "Believing as I do that man in the distant future will be a far more perfect creature than he now is, it is an intolerable thought that he and all other sentient beings are doomed to complete annihilation after such long-continued slow progress."

The grip of science was strong, though, and while these troubling debates and concerns continued, the twentieth century brought great improvements in the study of certain qualities of the Sun. Spectrum analysis allowed for study of the Sun's colored, infrared, and ultraviolet rays; improved instruments were capable of gaining better understandings about its structure, composition, and energy generation. Astronomers also discovered that sunspots were cyclical, with a duration averaging eleven years. An active body of solar physicists working in the 1860s and later dedicated themselves to expanding scientific knowledge of the Sun. They became accustomed to, and increasingly adept at, solar photography as a tool for studying the Sun. They discovered that the Sun has an atmosphere that rises above its visible surface, and noted relationships between activities on the Earth and

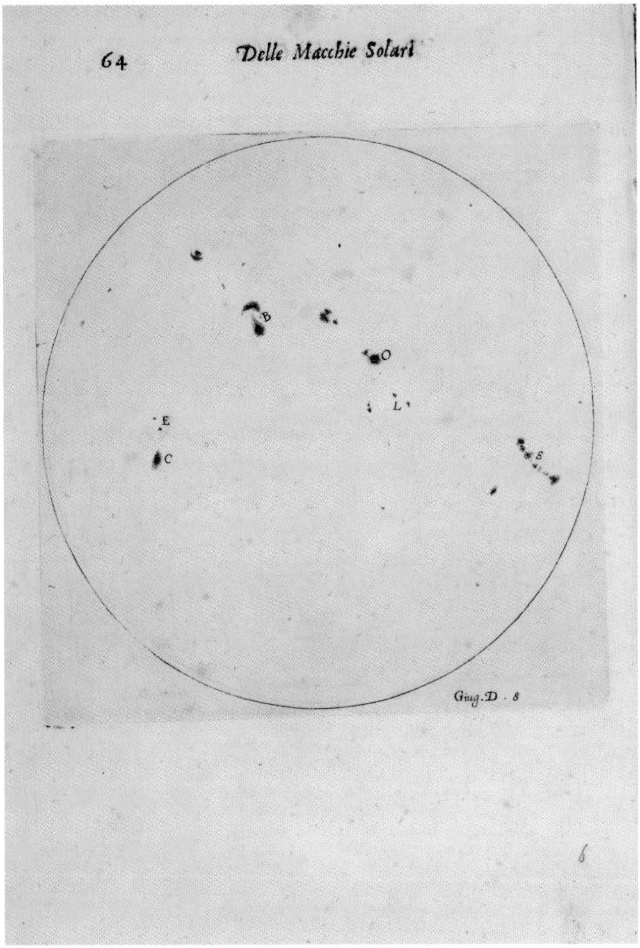

SOL
ELLI- PTIC:
HOC EST
Nouum & perpetuum Solis contra hi foliti Phæ- nomenon, quodnouiter inuentum,

STRENÆ LOCO,
REVERENDISSIMO ATQVE SERENISSIMO PRINCIPI AC DOMINO, DOMINO
MAXIMILIANO,
ARCHIDVCI AVSTRIÆ, DVCI BVRGVNDIÆ, STYRIAE, CARINTHIAE, CARNIOLAE ET WIRTENBERGAE; Magiſtratus Generalis Pruſſiæ Adminiſtratori; nec non Sanctæ militiæ Ordinisq; Teutonici per Germaniam, Italiam, partesq; transmarinas Magiſtro; Comiti Habſpurgi, Tyrolis & Goritiæ, Landgrauio Alſatiæ, &c. Domino ſuo clementiſſimo, offert, dicat, donat,

Chriſtophorus Scheiner Societatis IESV, *in Alma atque Catholica Ingolſtadienſi Academia, Sacræ Linguæ atque Matheſeos Profeſſor ordinarius.*

ANNO DOMINI CIↃ. IↃC. XV.

Cum facultate Superiorum.

AVGVSTÆ VINDELICORVM,
Typis Chriſtophori Mangij.

Ex Bibl. Jos. Ren. Card.

LEFT

Galileo Galilei. *Istoria e dimostrazioni intorno alle macchie solari e loro accidenti comprese in tre lettere...* (Rome: Giacomo Mascardi, 1613). This engraving shows significant sunspot activity in mid-August 1612. This edition, known as the "domestic" issue, contains Galileo's letters from archrival Christopher Scheiner's to Mark Welser, a scientific correspondent of both Galileo's and Scheiner's. Afraid of international copyright disputes, the publisher earmarked the copies including Scheiner's letters for distribution in Rome only. (Huntington Library: RB 83089)

RIGHT

Christopher Scheiner. *Sol Elliptic[us]* (1st edition. Augsburg: Christoph Mang, 1615). As early as 1612, Scheiner noticed that the Sun assumed an elliptical rather than circular shape as it approached the horizon. He explained it in terms of atmospheric refraction, a hypothesis he elaborated further in his *Refractions Coelestes* (1617) and continued later in his more detailed work on the Sun and sunspots, *Rosa Ursina*. (Huntington Library: RB 487000.83)

solar activity, such as weather patterns, changes in magnetic fields, and compositional similarities between the Sun and Earth.

These important activities and the new methods used to study the Sun led to still more discoveries and revelations in the twentieth century. One of the twentieth century's great statesmen of science, George Ellery Hale made vital contributions to solar astronomy. He founded the Mount Wilson Observatory in Southern California in 1904; invented the spectroheliograph, which took monochromatic images of the Sun; and built three superbly equipped telescopes for studying the Sun (the original purpose of an observatory on Mount Wilson was for solar study). Active in the creation of both the California Institute of Technology and the Huntington Library, he was an extraordinarily well organized and energetic man. He and his staff established the existence of strong magnetic fields in sunspots in 1908, and identified the twenty-two–year cycle of solar magnetic activity in 1924. Perhaps his crowning achievement was the initiation of construction of the 200-inch (five-meter) telescope at Mount Palomar, which is now called the Hale telescope in his honor.

Many historical facets of solar observation have yet to be successfully studied in depth. And despite sophisticated tools and the benefit of centuries of historical scrutiny and analysis, much modern work remains to be done on the most fundamental and life-giving component of our solar system. In 1934, Hale wrote to a colleague in England, "The fact is that we don't seem to *know* much about the Sun, try as we may." Many experts in the field would probably still agree. *DL*

Christopher Scheiner. *Rosa Ursina* (1st edition. Bracciani : Apud Andream Phaeum Typographum Ducalem, 1626–1630).

Right: The Duke of Orsini, who sponsored this work, is shown on the frontispiece surrounded by roses, the emblem of the Orsini family. The massive volume did not draw very positive reviews following its publication, nor did it go down well in history—primarily because Scheiner vigorously attacked his long-time nemesis Galileo in the first section of the book. Despite this diatribe, the work is important because it contains detailed sunspot data. Scheiner uses his observations of sunspot paths across the solar disk to demonstrate that the Sun's rotational axis is inclined with respect to the Earth's orbital plane.

Below: Scheiner tracked and described partial and total solar eclipses. His careful scrutiny of the Sun led him to several important discoveries, such as the Sun's rotational period (27 days), and the fact that sunspots existed only in what he called a "royal zone"—extending approximately thirty degrees on either side of the Sun's equator. (Huntington Library: RB 487000.1055)

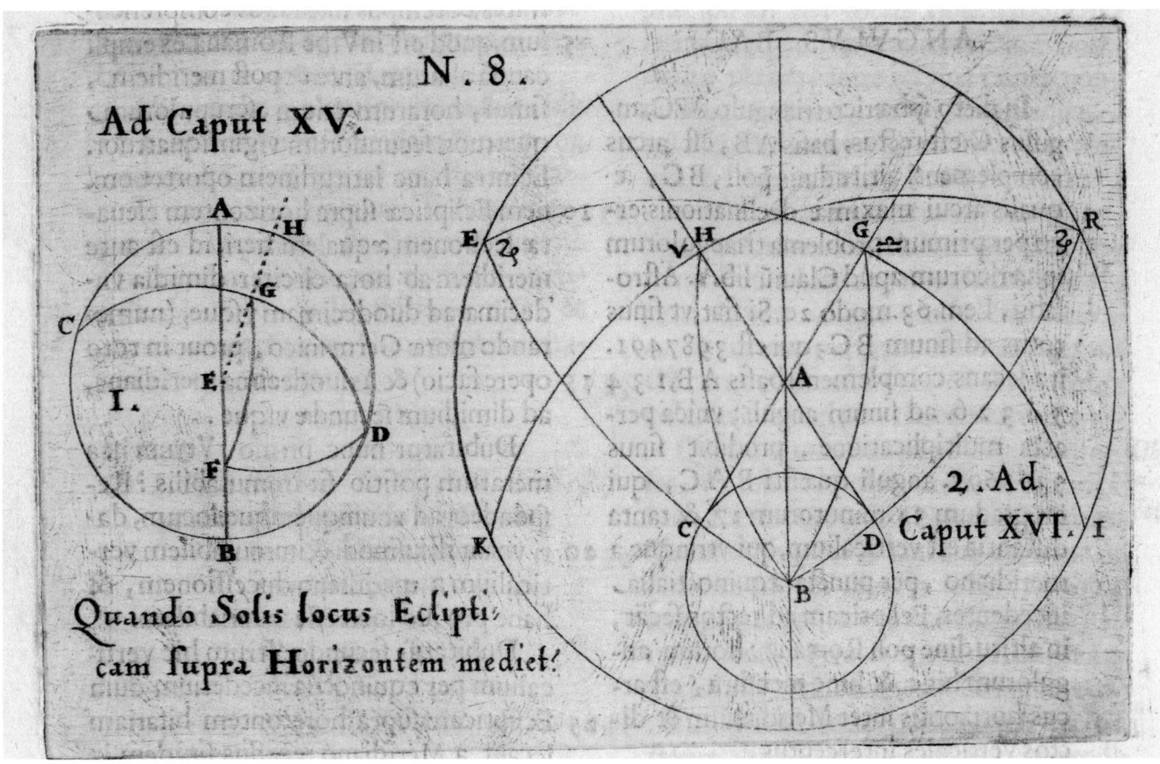

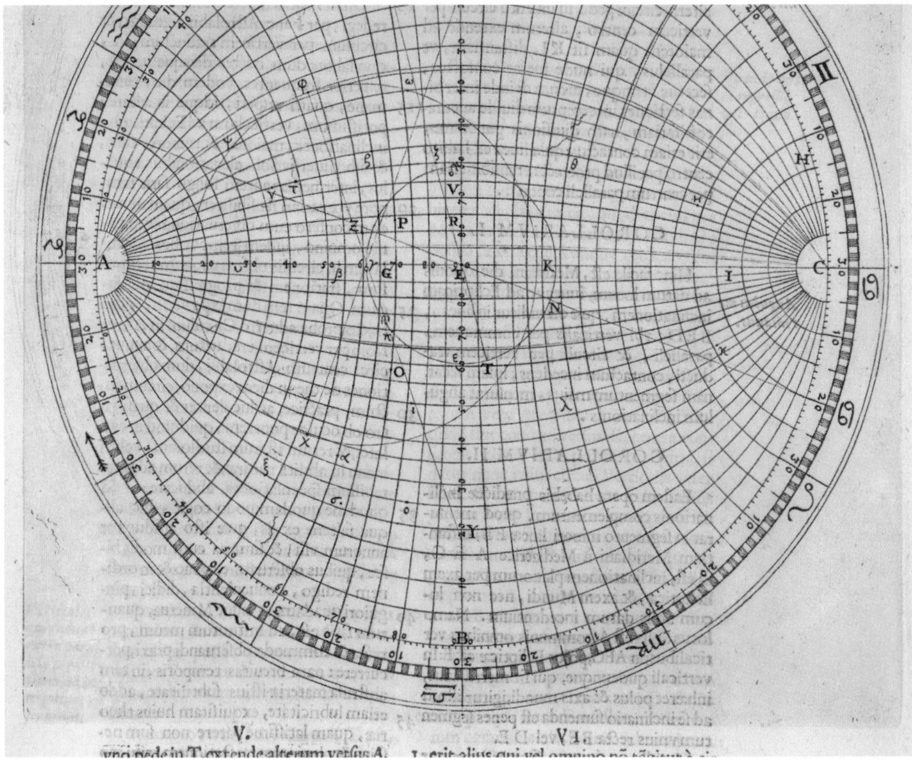

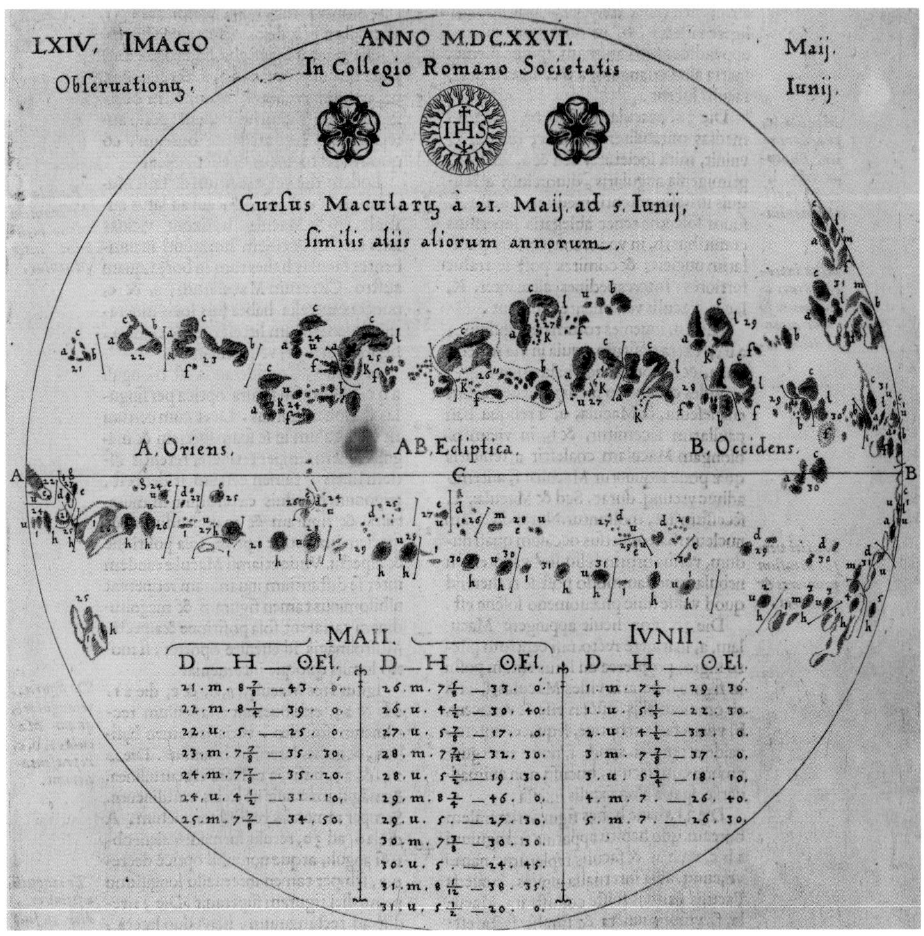

Christopher Scheiner. *Rosa Ursina* (1st edition. Bracciani: Apud Andream Phneum Typographum Ducalem, 1626–1630). Scheiner's method of illustrating the motion of individual spots across the face of the Sun became the standard method of describing the sunspots' motion and their changing shapes. His work proved extremely useful as an aid to studying sunspots because of a phenomena that followed the publication of the *Rosa Ursina*. Known as the Maunder Minimum, this period between 1645 and 1715 was one of unexpectedly and greatly reduced sunspot activity, and Scheiner's work provided recent comparative data with which to evaluate this "lull." (Huntington Library: RB 487000.1055)

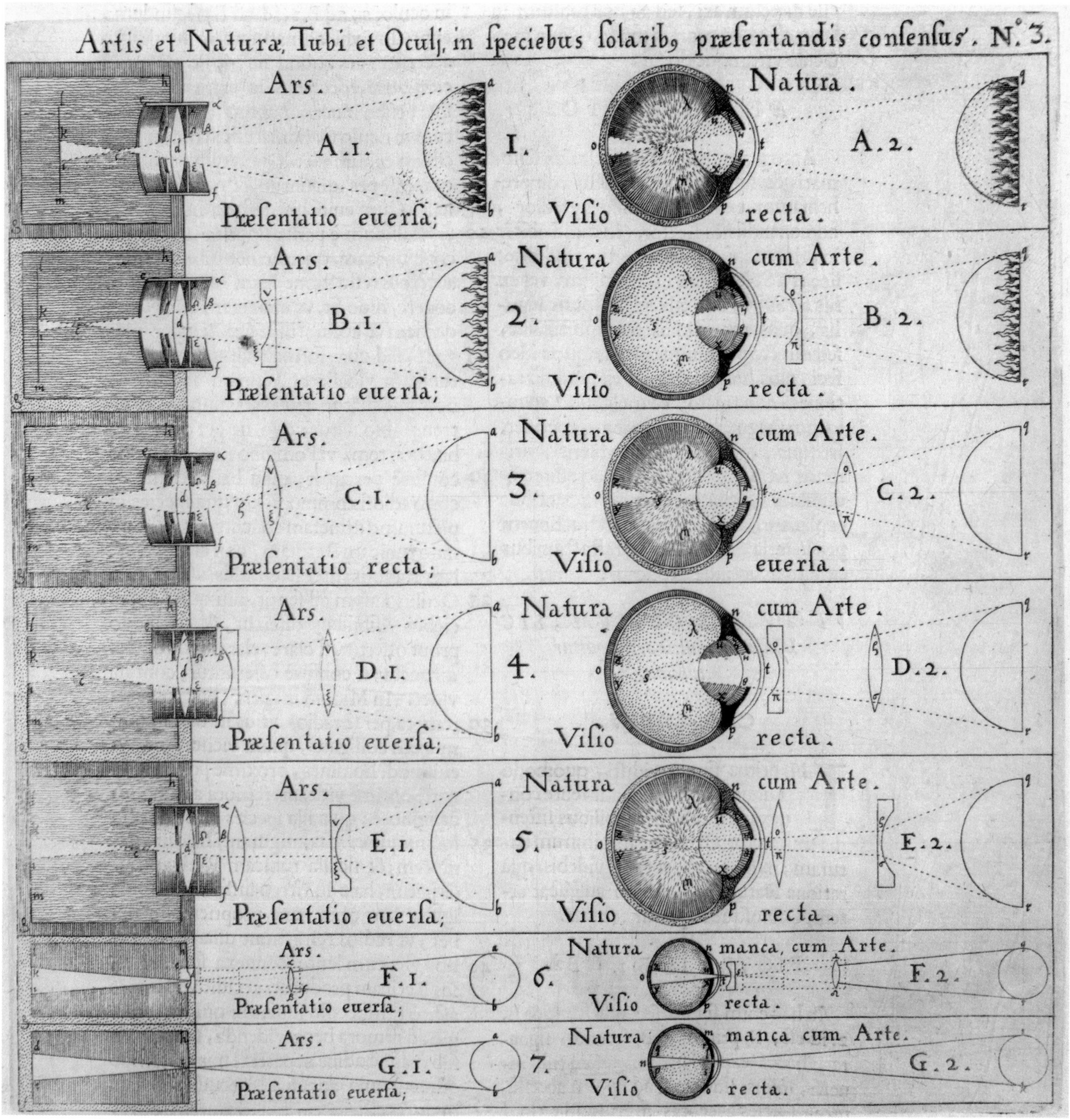

Christopher Scheiner. *Rosa Ursina* (1st edition. Bracciani: Apud Andream Phneum Typographum Ducalem, 1626–1630). Like Kepler and others before him, Scheiner was interested in the role of the eye as a part of the seeing equation: the lens, the Sun, and the eye all played crucial roles in the mystery of the Sun and its light. (Huntington Library: RB 487000.1055)

Christopher Scheiner. *Rosa Ursina* (1st edition. Bracciani: Apud Andream Phneum Typographum Ducalem, 1626–1630).

TOP: The solar observations of Scheiner were taken on several consecutive days by projecting an image of the Sun onto a sheet of white paper and tracing out the positions and appearance of sunspots. These observations, which detailed changes in the size and shape of individual sunspots from day to day, showed that the motion of sunspots across the disk of the Sun were due to the solar rotation.

BOTTOM: Scheiner's techniques for studying sunspots included enlisting several observers (primarily in Europe) to gather data on the days that he was unable to do so. In this sense, he served as a forerunner to modern solar astronomer George Ellery Hale's concept of placing several solar observatories in a logically spaced chain in order to obtain uninterrupted solar data.

OPPOSITE: Scheiner's "Helioscopium," used for his later sunspot observations. The device allowed for projection of the Sun's image for study and drawing with a minimum of eyesight strain or difficulty. (Huntington Library: RB 487000.1055)

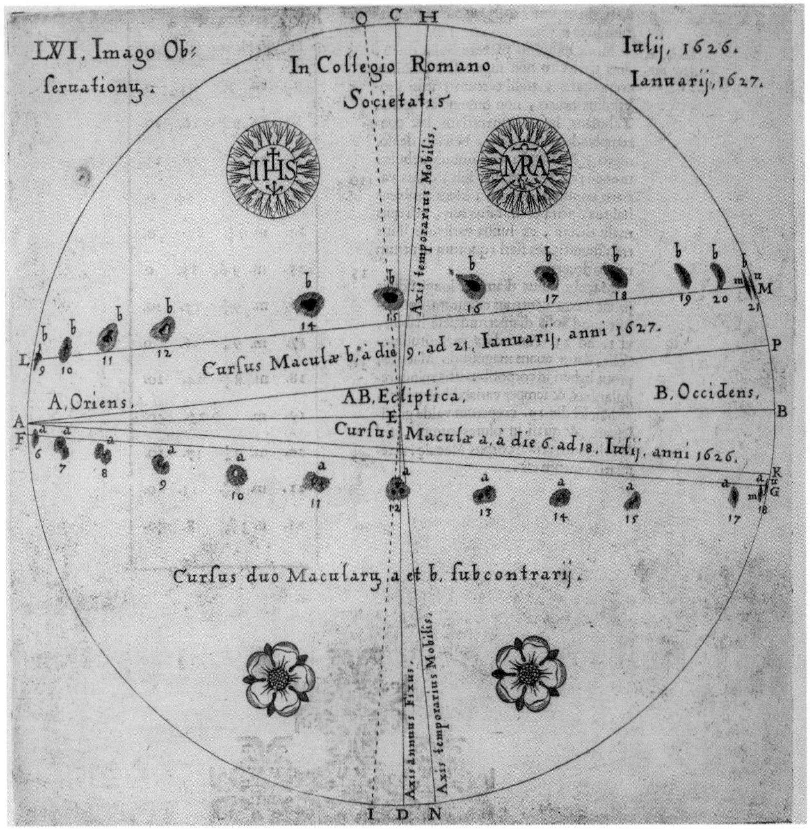

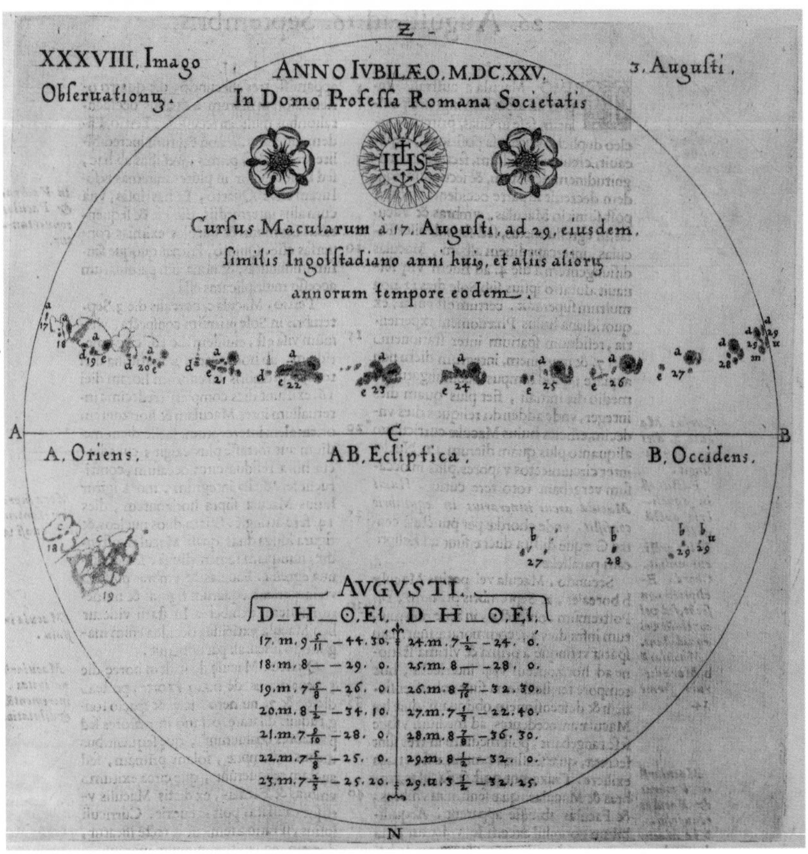

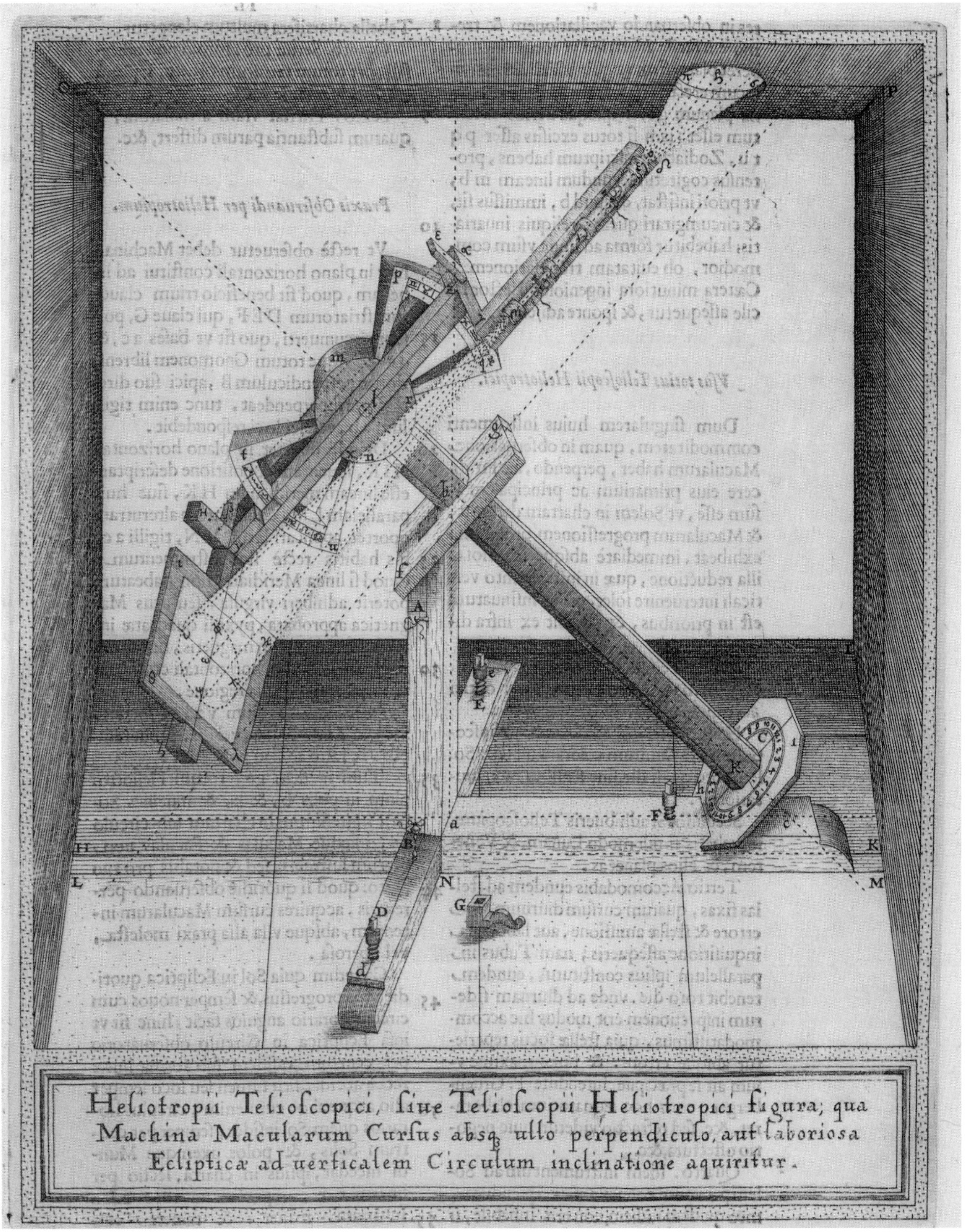

Heliotropii Telioscopici, siue Telioscopii Heliotropici figura; qua Machina Macularum Cursus absq; ullo perpendiculo, aut laboriosa Ecliptica ad uerticalem Circulum inclinatione aquiritur.

TABVLA SELENOGRAPHICA
Seu
Vera adminiculo tubi. orthograph.
delineatio, Marium Sinuum Insularum
Continent. Promont. Lacuum Paludum
Montium Planit. Vallium in visibili
Lunæ hemisphærio existentium
Autore
Johanne Hevelio
Aº 1645
EXPLICATIO LITERARVM
M. Mons. I. Insula.
S. Sinus. L. Lacus.
P. Palus. C. Caput.
Pr. Promontorium.
Pal. Paludes Fl. Fluvius
Fret. vel F. Fretum
REGIO HYPERBOREA
MARE HYPERBOREVM
Sinus Hyperb.
Paludes Orient.
MARE MEDITERR.
SICILIA
PONTVS
PROPONTIS
MARE ADRIATICVM
PALVS MOEOTIS
AEVXINVS
CHERSONESVS
Paludes Hyperb.
MARE EOVM
Syrticum Mare
Mare Ægyptiacum
Mare Pamphilium
Cassiotis Regio
S. Sirbonis
COLCHIS
MARE CASPIVM
ASIA
PERSIA
Desertum Mingui
Scala Milliarium Germanicorum
Digiti Ecliptici et eorum Segmenta
Fig. 8

THE MOON

OPPOSITE
Johannes Hevelius. *Selenographia, sive Lunae descriptio, atque accurata tam macularum ejus...* (Gdansk: Hünefeld, 1647). The *Selenographia* was the first in a series of books by Hevelius that are true masterpieces of seventeenth-century engraving and printing. (Huntington Library: RB 487000.1003)

THE MOON, CLOSEST CELESTIAL BODY TO EARTH, has captivated scientists, artists, and dreamers since antiquity.[9] "How sweet the moonlight sleeps upon this bank!" enthused Lorenzo in Shakespeare's *The Merchant of Venice.* First believed to consist of a smooth and unbroken sphere, the Moon quickly proved, with the invention of the telescope, to have a surface pockmarked with hundreds of thousands of craters, ragged edges, and uneven coloring. The ability after 1610 to observe these elements telescopically helped to dismantle thirteen hundred years of Aristotelian throught about the "perfect nature" of the heavens. A sharper view of the Moon only raised more questions: How were the craters formed? Did they stem from volcanoes, or were they the result of titanic collisions of foreign bodies? Was the surface covered in dust for many meters, or was it solid? These questions stood until the twentieth century, and some aspects of each of these questions remain partially unanswered.

Much like observers of the further stars and their constellations, Moon viewers could find any number of different objects among its light and dark patterns, including the "man in the Moon," as well as other items of folklore dating back through the centuries. In the late twelfth and early thirteenth centuries, the German philosopher Albertus Magnus (1193–1280) described a grouping of a dragon, a tree, and a little man in the Moon. Some 350 years later these were evoked by Shakespeare in the slightly changed form of a man with a dog, a bush, and a lantern.

Many observers before the advent of the telescope considered the Moon to be habitable. Some also speculated that the dark spots on the Moon were lunar seas. Only after the Apollo missions was it confirmed that they were relatively recent basaltic lava flows that had covered up some of the Moon's craters and left a smooth surface. These features are still described as seas (or lakes or basins or other liquid formations), a legacy that is also reflected in the Moon's formal nomenclature: the Sea of Tranquility, for instance.

The historiography of the Moon begins with the publication of Galileo's *Sidereus Nuncius* (Starry messenger) in 1610. His illustrations of the Moon have received adverse responses from shortly after their publication up into the twentieth century; but this criticism is misplaced and owes largely to extraordinary deterioration in the quality of images reproduced in later editions of his collected works, all which have included the *Sidereus Nuncius.* The five copperplate images as printed in the 1610 edition have to be seen to believed; despite their simplicity relative to modern illustrations, their crispness repudiates later editions and reproductions and exonerates Galileo of creating images lacking in the detail expected of a scientist. His 1610 images were often reproduced, but even subtle changes to the images eliminated important details. The images show four phases of the Moon—

James Nasmyth and James Carpenter. *The Moon: Considered as a Planet, a World, and a Satellite* (1st edition. London: John Murray, 1874). Among the highlights in this book is Nasmyth's "wrinkle theory"—as an object ages and loses its vital heat, all of it, except its skin, shrinks, making the skin pucker, wrinkle, and shrivel. (Huntington Library: RB 487000.927)

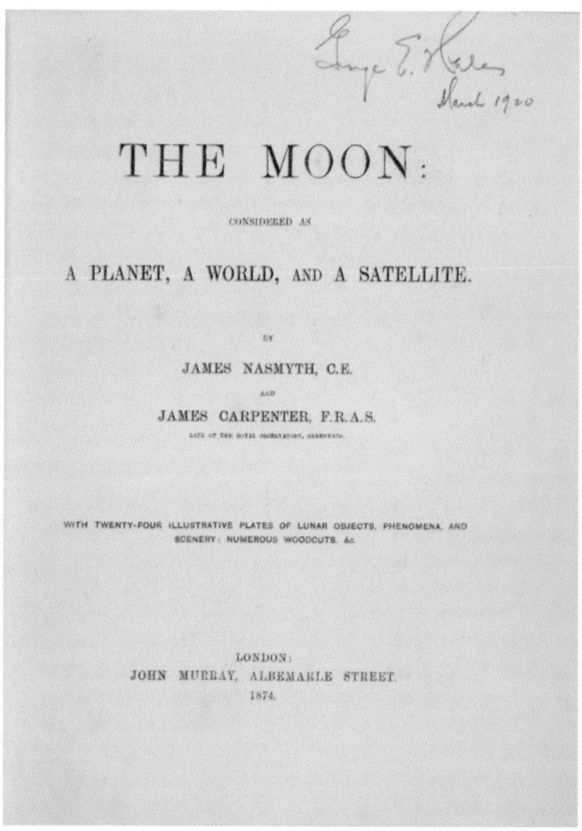
THE MOON:

CONSIDERED AS

A PLANET, A WORLD, AND A SATELLITE.

BY

JAMES NASMYTH, C.E.

AND

JAMES CARPENTER, F.R.A.S.

WITH TWENTY-FOUR ILLUSTRATIVE PLATES OF LUNAR OBJECTS, PHENOMENA, AND SCENERY; NUMEROUS WOODCUTS, &c.

LONDON:
JOHN MURRAY, ALBEMARLE STREET.
1874.

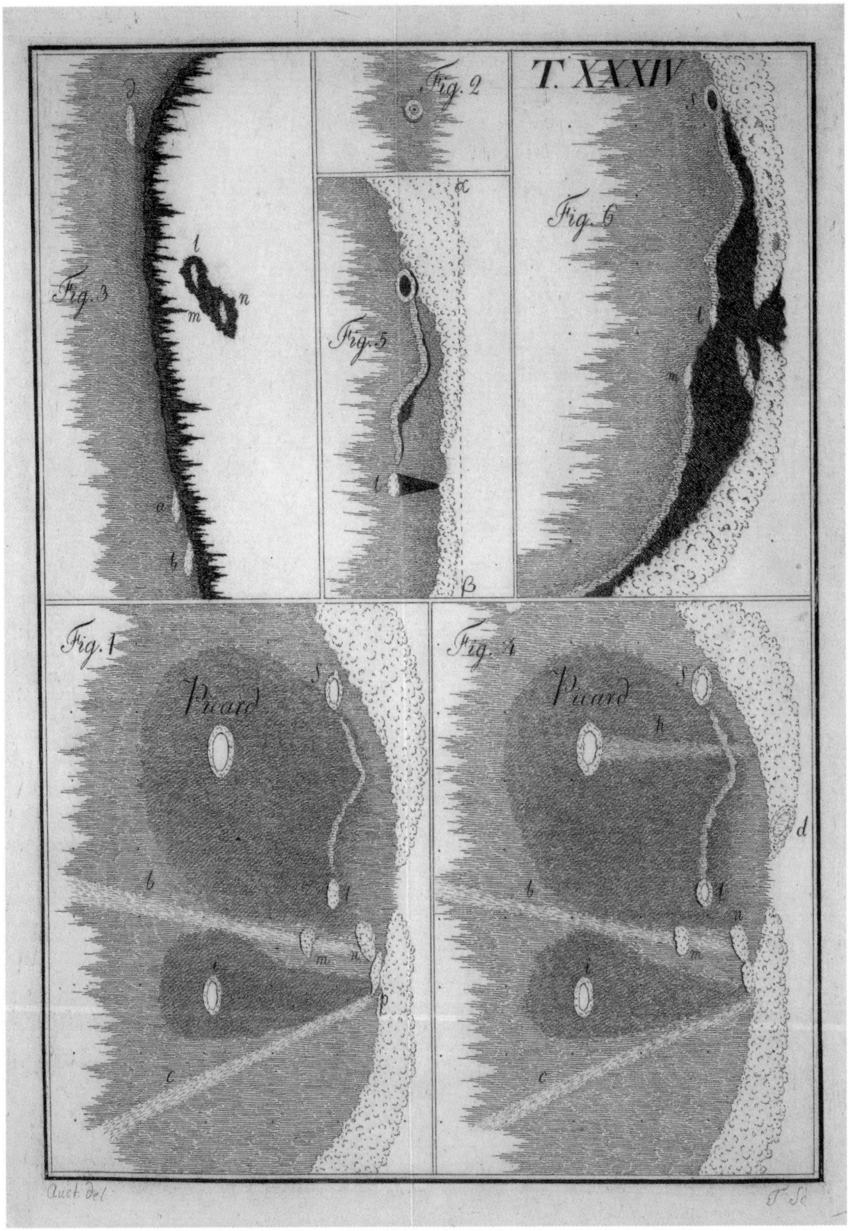

Johann Schroeter. *Selentopographische fragment...* (Goettingen: J.G. Rosenbusch, 1791). Bird's-eye view of different lunar craters. Schroeter, trained as a lawyer in Germany, developed an extracurricular interest in mathematics and astronomy. He closely studied lunar topography, creating detailed maps of the lunar surface by using the monster telescopes in use in the last quarter of the eighteenth century. (Huntington Library: RB 492372)

the crescent phase, the first quarter, the waning gibbous phase, and the last quarter (shown twice). The book shows no images of the full Moon because Galileo wanted to illustrate the Moon's substantial roughness, which is less visible in the full phase.

The next important author to illustrate and write about the Moon was Johannes Hevelius, who published his *Selenographia* in 1647. The first edition became one of the monuments of Renaissance astronomy, the first large and widely available lunar atlas, and a visually and artistically striking work that received extraordinary acclaim throughout Europe. The *Selenographia* is one of the finest scientific books of the seventeenth century; the original and valuable information it contains was unsurpassed for many years. The first chapters deal with Hevelius's method of producing high-quality lenses and instruments. Hevelius—like Tycho Brahe, most notably—invented and constructed very precise instruments, including telescopes of various sizes. For his lunar observations, he used various telescopes, including a

Johannes Hevelius. *Selenographia, sive Lunae descriptio, atque accurata tam macularum ejus...* (Gdansk: Hünefeld, 1647). The value of the *Selenographia* may be judged not so much by the praise of Hevelius's contemporaries as by the fact that even in that time of rapid development in astronomy, his lunar maps remained among the finest and most heavily consulted for more than a hundred years. (Huntington Library: RB 487000.1003)

very large one, but for research of the stars he strongly opposed them, advocating instead precision sights which he used as accurately as others used optics.

Hevelius engraved the illustrations for the *Selenographia* and published the work himself in Gdansk. There are 111 fine full-page engravings and twenty-six additional ones in the text, including a series depicting the lunar phases day by day. The *Selenographia* was the first in a series of books by Hevelius that are masterpieces of seventeenth-century engraving and printing. It is likely that while Hevelius was a student in Leyden, he became acquainted with the fine printing techniques of the Elsevier family, for the *Selenographia* is printed in the best Elsevier folio tradition.

By the early nineteenth century, driven by advances in telescopes, lunar cartography evolved to the point where constructing a large-scale map of the Moon on a par with maps of the Earth became truly possible for the first time. The Dresden surveyor/cartographer, William Lohrmann (1796–1840), constructed a large-scale

Johannes Hevelius. *Selenographia, sive Lunae descriptio, atque accurata tam macularum ejus...* (Gdansk: Hünefeld, 1647). The *Selenographia* received extraordinary acclaim. In Paris, astronomers who received gift copies from Hevelius had little opportunity to read it because of the friends and acquaintances who came to see and marvel over the book. The Pope himself saw a copy and said it would be a book without parallel, had it not been written by a heretic. (Huntington Library: RB 487000.1003)

map in twenty-five square sections, with detailed and meticulous rendering of lunar features. Because of the increased scale and detail of his map, Lohrmann found it necessary to use a numbering or lettering schema to identify individual elements. His effort was one of the first geodetic surveys of the Moon, designed to provide unprecedented levels of accuracy. Some features in his survey were named, and others were simply numbered. Lohrmann died before completing the work, but the effort was taken up by others, most notably amateur astronomers Wilhelm Beer (1797–1850) and Johann Heinrich Mädler (1794–1874). They published their work, which owed a large debt to Lohrmann, in four sections between 1834 and 1836. They left the impression that there was little left to achive in lunar studies, and virtually no serious work was done to map the Moon for the next thirty years, until lunar photography had reached a state that allowed for still more advanced and detailed work.

Photography altered our perception of the physical world in the nineteenth century, changes that are reflected in both scientific thought and illustration. One of the first photographs ever taken was an image of the Moon taken by Louis Daguerre in 1838. Celestial photography allowed people to view the Moon clearly and to have a supremely accurate printed record of the Moon during its different phases. Pioneering advances in lunar photography were made by such important figures as the wealthy amateur astrophotographer Warren De la Rue at mid-century. Lunar photography allowed for a number of advances. The Moon's details could be recorded in great detail in an instant, making problems with illustration

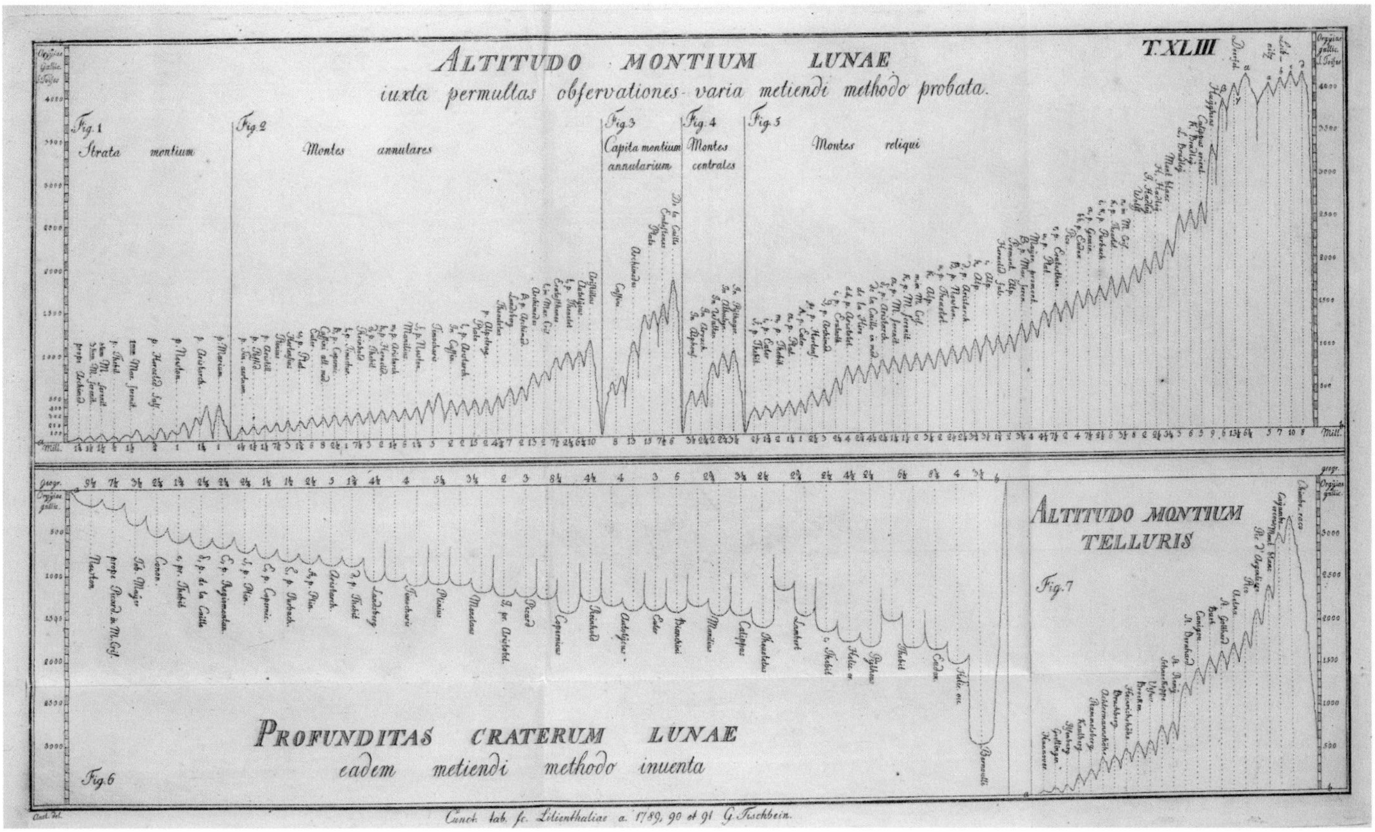

Johann Schroeter. *Selentopographische fragment...* (Goettingen: J.G. Rosenbusch, 1791). Lunar mountains and craters shown in profile. Schroeter's two volumes on lunar theory provided much new information about moon craters in profile. Although he did not use the most precise measuring devices available, such as a micrometer, he nevertheless took painstaking measurements and identified a number of lunar formations for the first time. (Huntington Library: RB 492372)

and related issues of changing light, cloud cover, and movement in the heavens moot. The precision of photographic machinery allowed for reproduction of segments of the Moon at precisely the same magnification, allowing segments of the Moon, photographed in fine detail, to be lined up and reproduced in large atlases. Photography also allowed people with no artistic skill to record images for scrutiny.

The first lunar atlases to include photographs did not include photos of the Moon itself. The Englishmen James Nasmyth (1808–1890) and James Carpenter (1840–1899) constructed plaster models of the Moon based on telescopic drawings. Photos of the models were published as *The Moon, Considered as a Planet, a World, and a Satellite,* in 1874. Nasmyth and Carpenter provided not only physical descriptions of lunar features, but tried to describe the origins of the features as well.

Many mysteries still surround the Moon, including a number that have been debated into the late twentieth century, such as the origin of the Moon itself. For instance, it was not until after the lunar missions of the 1960s and 1970s that scientists agreed that the craters of the Moon were caused by massive impacts and not by volcanic activity. (The aforementioned basaltic lava flows that followed the creation of these ancient craters was volcanic, but the craters themselves were not.) Other questions are still viable topics of debate and discussion: How was the Moon itself formed? Was it a body traveling though space captured by Earth's gravity? Was it formed from the Earth itself? The latest theory, one with considerable assent among astronomers, was proposed in October 1984 at a conference in Kona, Hawaii: a Mars-sized object struck the Earth while it was still in a molten state but after its

James Nasmyth and James Carpenter. *The Moon: Considered as a Planet, a World, and a Satellite* (1st edition. London: John Murray, 1874).
RIGHT: The images from the Huntington's copy of this work, once owned by George Ellery Hale, show the Moon's craters, believed by Nasmyth and Carpenter to be of volcanic origin. This image shows the ejecta matter they believed was formed by the lunar eruption.

LEFT: The formation of a volcanic crater is shown, "when the principal cone was vomiting forth ashes, stones, and red-hot lava." (Huntington Library: RB 487000.927)

this excavating of the hollow and widening of its mouth and mound would be extended. But when a weaker outburst came, or when the energy of the last eruption died away, a process of slow piling up of

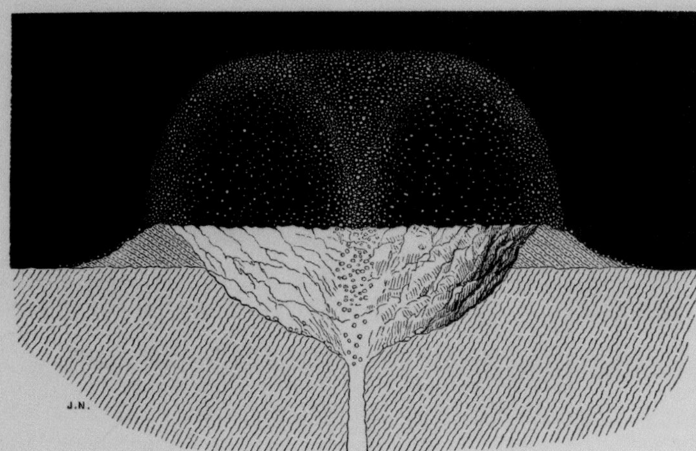

FIG. 23.

matter close around the vent would ensue. It is obvious that when the ejective force could no longer exert itself to a great distance it must merely have lifted its burden to the relieving vent and dropped it in the immediate neighbourhood. Even if the force were considerable, the effect, so long as it was insufficient to throw the ejecta beyond the rim of the crater, would be to pile material in the lowermost part of the cavity; for what was not cast over the edge would roll or flow down the inner slope and accumulate at the bottom. And as the eruption died away, it would add little by little to the heap, each expiring effort leaving the out-given matter nearer the orifice, and thus building up the central cone that is so conspicuous a feature in terrestrial volcanoes, and which is also a marked one in a very large proportion of the craters of the moon. This formation of the cone is pictorially described by Fig. 24.

In the volcanoes of the earth we observe another action either concurrent with or immediately subsequent to the erection or formation of the cone: this is the outflow or the welling forth of fluid lava, which

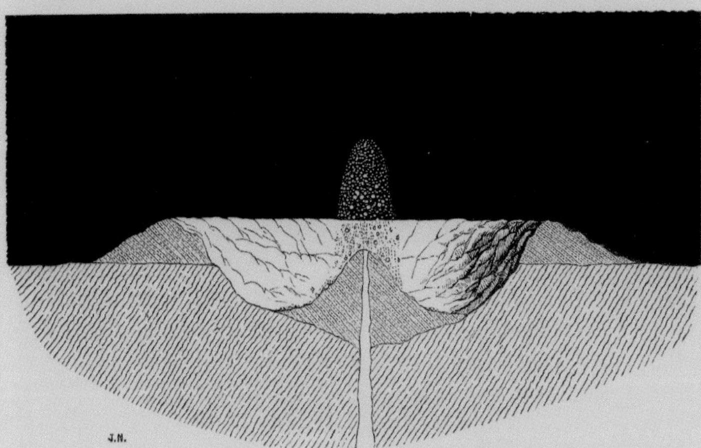

FIG. 24.

in cooling forms the well-known plateau. We have this feature copiously represented upon the moon and it is presumable that it has in general been produced in a manner analogous to its counterparts upon the earth. We may conceive that the fluid matter was either spirted forth with the solid or semisolid constituents of the cone, in which case it would drain down and fill the bottom of the crater; or we may suppose that it issued from the summit of the cone and ran down its sides, or that, as we see upon the earth, it found its escape before reaching the apex, by forcing its way through the basal parts. These actions are indicated hypothetically for the moon in Fig. 25; and the parallel phenomena for the earth are shewn by the actual case (represented in Fig. 26 and on Plate I.) of Vesuvius as it was seen by one of the authors in 1865, when the principal cone was vomiting forth ashes, stones, and red-hot lava, while a vent at the side emitted very fluid lava which was settling down and forming the plateau.

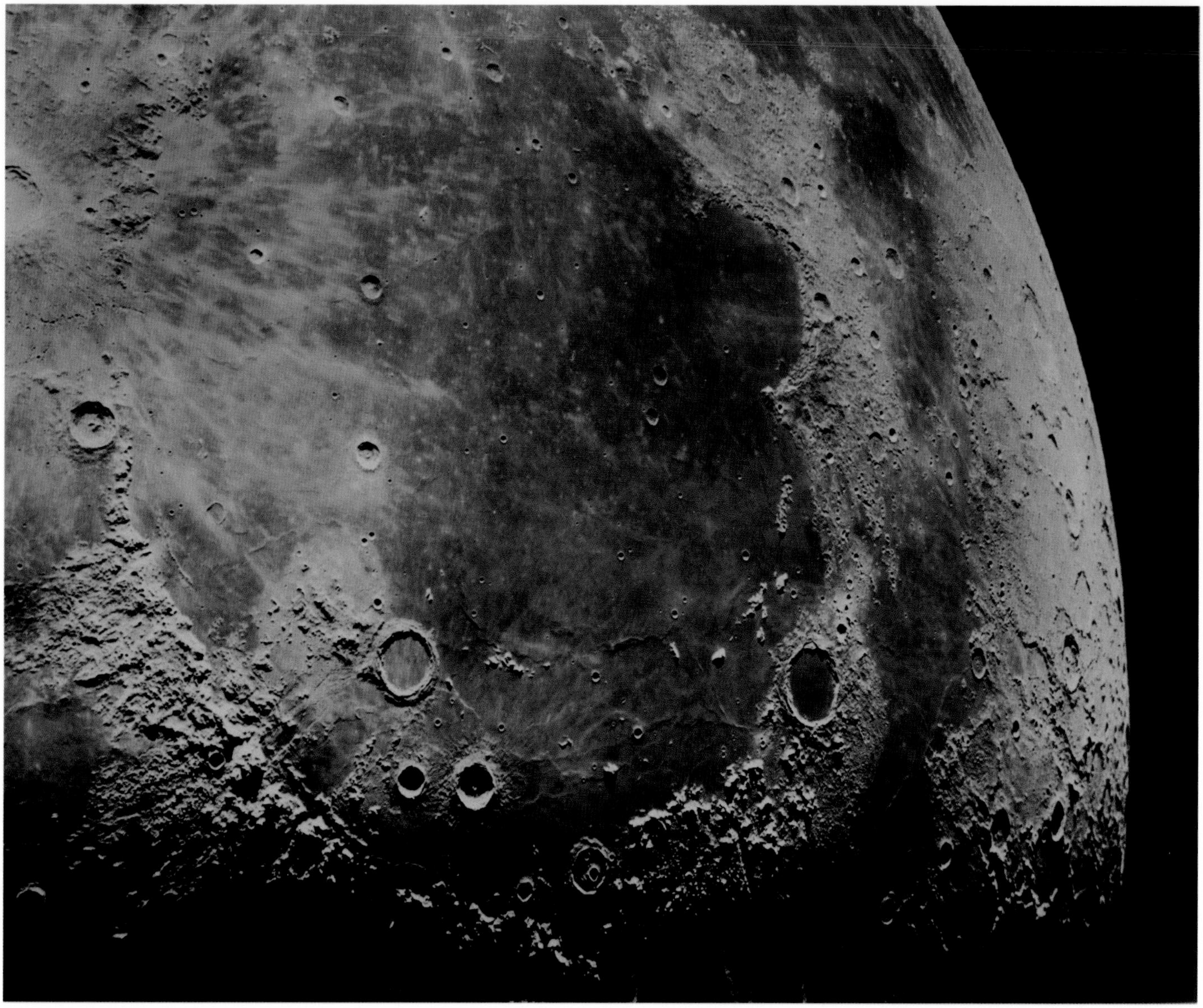

Twentieth-century photograph of the Moon, showing the Apennines, Alps, Plato, Copernicus, and Mare Imbrium regions, may have been taken at Mount Wilson near Pasadena. The relatively smooth dark maria, or "seas," cover about 16 percent of the lunar surface. The relatively bright and more heavily cratered highlands are called terrae. The craters and basins in the highlands are formed by meteorite impact and are thus older than the maria, having accumulated more craters. (Huntington Library: Photo Collection, photOV 10170)

OPPOSITE
This image was taken by the *Clementine* spacecraft on March 13, 1994 as it came over the north lunar pole at the completion of a mapping run. The large crater at the bottom of the image is Plaskett (180 W longitude, 82 N latitude). (National Space Science Data Center, NASA)

heavy core had separated and begun to sink toward the center. The core of the "proto-earth" and the impactor merged, and a mix of lighter mantle material flung off from both bodies eventually accreted into what is now the Moon. The moon programs of the twentieth century have allowed for actual physical study of the Moon's material. Between 1969 and 1972, six Apollo space flight missions brought back 382 kilograms (842 pounds) of lunar rocks, core samples, pebbles, sand, and dust from the lunar surface. In addition, three automated Soviet spacecraft returned important samples totaling 300 grams (approximately 3/4 pound) from three other lunar sites. Johnson Space Center is the chief repository for the Apollo samples. The lunar sample laboratory at the Johnson Space Center in Houston prepares pristine lunar samples prepared for shipment to scientists and educators. Nearly a thousand samples are distributed each year for research and teaching projects. The rock on display at the Huntington exhibition comes from this NASA program. *DL*

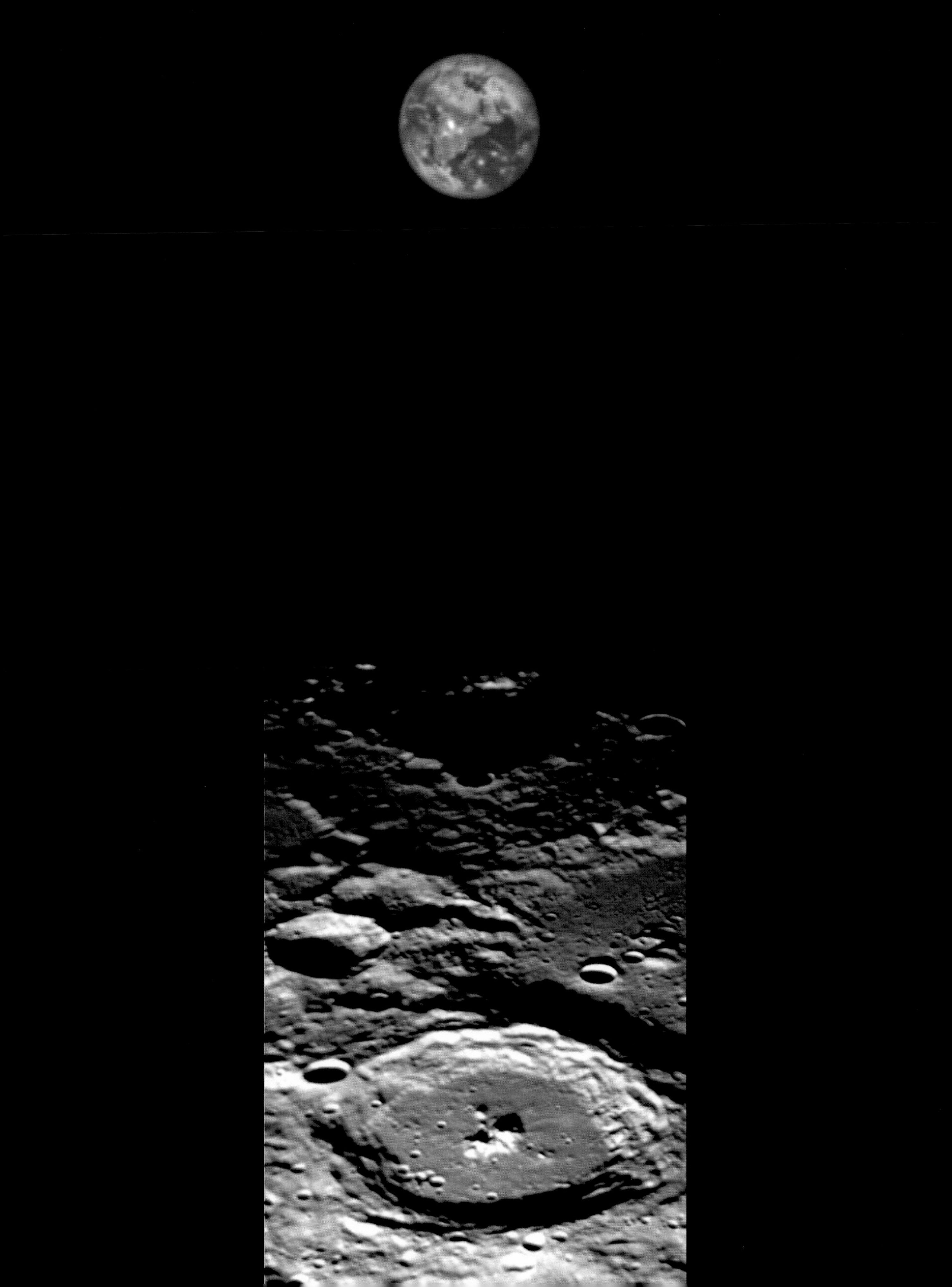

FIFTEEN OTHER PLANETS

Saturn color-enhanced with rings. (NASA/JPL Image)

WHEN THE ANCIENT PEOPLES GAZED AT THE SKY, they would easily notice the Sun, Moon, and the fixed stars. But among them were five bodies that appeared to be very bright stars but which moved slowly against the background of fixed stars. These bodies were called *planetes* ("wanderers") by the Greeks. The planets played key roles in the study of astrology and were ascribed great power in influencing human behavior. They were quite revered by many civilizations, often seen as manifestations of gods; we currently use the Roman names for them. As they moved across the sky, the five planets would generally drift from west to east against the fixed stars, but occasionally they would stop, reverse direction for a short time, stop again, and return to their eastward path. Their motion was contained to a very narrow path across the sky, which also marked the travels of the Sun and the Moon, that came to be known as the ecliptic. The ancient and medieval astronomers struggled to explain the unusual behavior of the planets within the context of perfectly circular motion, but never quite achieved success despite their best efforts.

Mercury, the nearest planet to the Sun, was always difficult to observe because it rarely moved far enough away from our star's blinding light. A number of early astronomers thought they caught a glimpse of an atmosphere around the planet and others searched for suggestions of surface markings, hoping to determine how long a Mercurian day was. For a long time the accepted notion held that Mercury always kept the same side toward the Sun, in the same way that the Moon always shows only one hemisphere to us. Radar observations in 1965 using the Doppler effect, revealed that it took nearly fifty-nine days for Mercury to complete one rotation on its axis. The spacecraft *Mariner 10* observed Mercury in the early 1970s, finally showing it to be a lifeless, cratered world. *Mariner 10* was the only spacecraft to ever explore Mercury, and so we still know very little about it. What little data we have about Mercury indicate that it is the densest planet, that its surface is very old and may reveal clues about the early solar system, that it has a thin exotic atmosphere, that it has a global magnetic field, that its temperatures vary from the highest to lowest temperatures in the solar system, and that ice deposits may exist at its poles. In order to learn more about these mysteries, a new mission, called *Messenger*, is planned to launch in 2004, fly by Venus twice and Mercury twice, before entering orbit around Mercury in 2009.

Unlike Mercury, Venus was always fascinating to observe, the brightest object in the heavens save for the Sun, Moon, and occasional comet or supernova. But like Mercury, Venus never strayed too far from the Sun, appearing at its best just before sunrise or just after sunset. In 1610 Galileo claimed to have seen Venus exhibit phases just as the Moon does; this was a major blow to the Ptolemaic concept of an Earth-centered universe because Venus could only exhibit a complete cycle of phases if it orbits the Sun. Astronomers peered with great effort at Venus, trying to find a glimpse of surface markings and so calculate the length of the Venusian day. Francesco Fontana (1575–1656), in his *Novae coelestium, terrestrium*

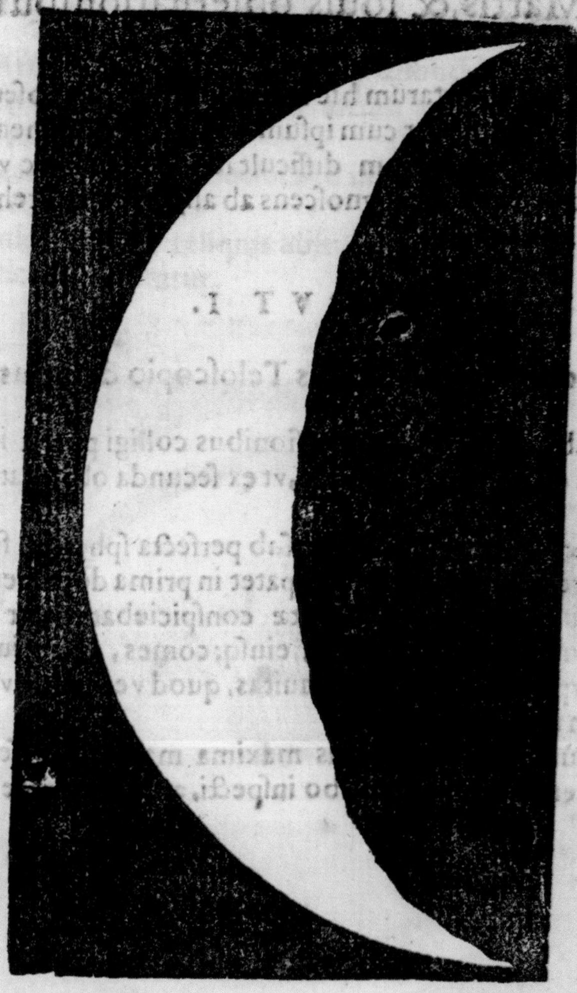

OB-

Simon Marius. *Mundus Iovialis*... (Noribergensis : Sumptibus Ex typis I. Lauri, 1614). This image of Marius includes a telescope at the bottom and at the upper left a depiction of the orbits of the four large moons of Jupiter. (Huntington Library: RB 481440)

OPPOSITE

Francesco Fontana. *Novae coelestium*... (Napoli: Apud Gaffarum, 1646). The crescent of the planet Venus as seen through one of Fontana's telescopes. (Huntington Library: RB 487000.420)

rerum observationes (The new heavens as observed from Earth) of 1646, claimed to have seen very faint markings on the planet and estimated a day on Venus to be much like the Earth: 23 hours and 41 minutes. Fontana would have been chagrined to find out that Venus makes one complete rotation on its axis in 243 days (although this was determined finally in 1964)! During the very rare transits of Venus across the face of the Sun in the eighteenth century, some astronomers found evidence of an atmosphere around Venus—this was finally confirmed two centuries later. As *Mariner* 2 passed by the planet in 1962, it became clear that Venus would be a place inhospitable to life as we know it. Later spacecraft confirmed this view by detecting surface temperatures of nearly 500 degrees Celsius and atmospheric pressures close to one hundred times that on the Earth's surface. One very puzzling aspect regarding Venus is why it has evolved into a much different planet than the Earth. Venus and the Earth are nearly the same size, but Venus has a dense atmosphere of carbon dioxide, which has produced a runaway greenhouse effect causing the extremely high temperatures. Venus has been explored in great detail since the 1970s, and the recent *Magellan* mission found substantial evidence of recent geological activity including lava flows and quake faults.

Mars, identified by its reddish color with the Roman god of war, played a key role in the success of the Copernican system. Johannes Kepler, using the detailed observations of Mars done at Tycho Brahe's observatory, plotted out the Martian orbit and came to the conclusion that it was an ellipse and not a perfect circle. Unlike Mercury and Venus, Mars presented obvious surface markings, including two polar ice caps. In the nineteenth century, the two moons of Mars were discovered, and reports came out about the appearance of canals on the Martian surface. Since the 1960s Mars has been the target of a large number of spacecraft, which showed us that Mars was canal-free and cratered, with evidence of past geological activity and apparently lifeless. In the year 2000 the *Mars Global Surveyor* spacecraft found evidence that liquid water has seeped onto the Martian surface in the geologically recent past. Future Mars missions, including *Mars Surveyor 2001*, will continue to examine the planet in great detail.

Jupiter, by far the largest of the planets in our solar system, was a natural target for Galileo's telescope in 1610. He quickly discovered four moons orbiting the planet, which demonstrated that not all orbital motion in the universe need be around a central Earth, further weakening Ptolemy's argument about the structure of the universe. Galileo named the moons for a potential patron, the Duke of Medici, but the names that we now use for the moons—Ganymede, Callisto, Europa, and Io—were provided by Galileo's rival, Simon Marius (1573–1624), in his book, *Mundus Iovialis* (The Jovian worlds) of 1614. We note Galileo's importance to these moons by referring to them collectively as the Galilean satellites. Later in the seventeenth century, astronomers noted a large red spot on the planet's surface, an atmospheric storm that has persisted until this day and is now known by the highly technical name of "The Great Red Spot." Jupiter itself is known as a "gas giant," a planet composed primarily of gases with small rocky cores. In the 1970s, four space-

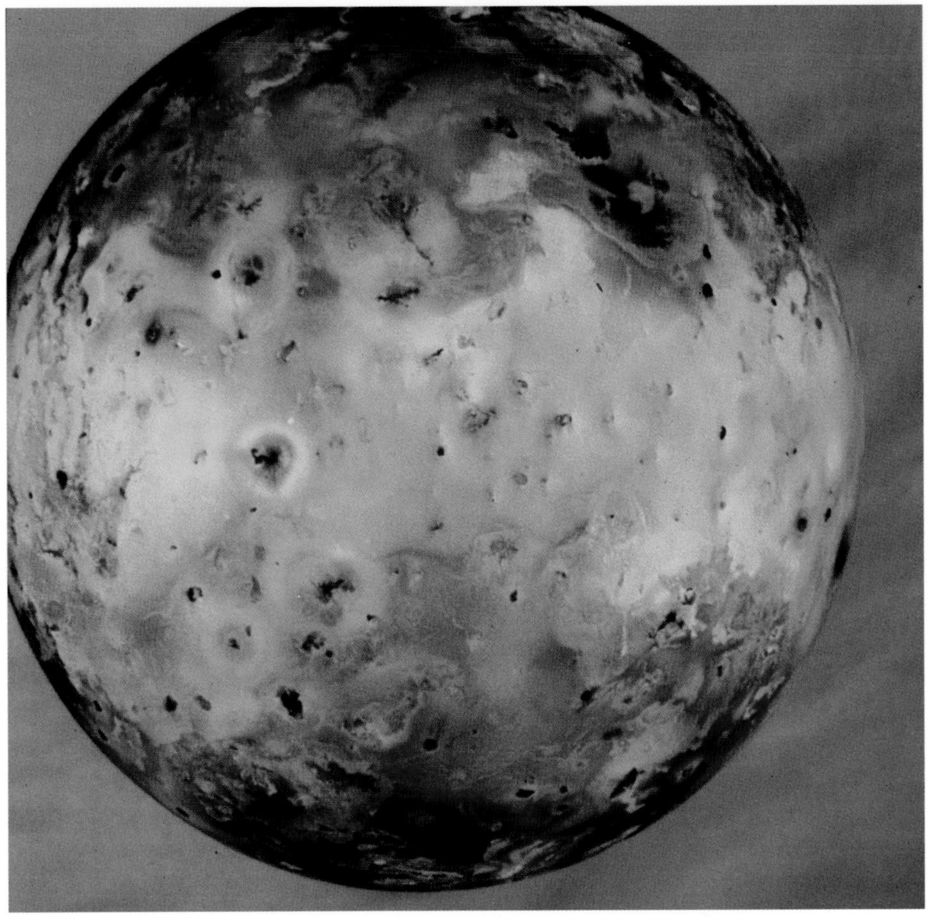

Io, the most volcanic body in the solar system, seen in front of Jupiter's cloudy atmosphere. The Galileo spacecraft acquired its highest resolution images of Io on 3 July 1999. This false-color mosaic uses the camera's near-infrared, green, and violet filters (slightly more than the visible range), processed to enhance more subtle color variations. Most of Io's surface has pastel colors, punctuated by black, brown, green, orange, and red areas near the active volcanic centers. (NASA)

craft flew by Jupiter providing us with a great deal of additional information about the planet and its moons (now known to number sixteen), including the fact that the planet is surrounded by a faint, narrow ring of debris. In 1995 an orbiting mission, *Galileo*, revealed such aspects of the Jovian system as the existence of hot, active volcanos on Io (first discovered by the *Voyager* spacecraft) and the possibility that large oceans may exist under the surfaces of Europa and Callisto.

The last of the five planets visible to the naked eye, Saturn, provided a quite unusual appearance to the astronomers who first turned telescopes toward it. Saturn looked to have some sort of protrusions extending out from either side of the planet. About forty-five years later, Christiaan Huygens realized that there was a large ring circling the planet. Huygens also discovered a moon (later named Titan) orbiting Saturn. Saturn was found to be, like Jupiter, a gas giant with a system of rings and satellites surrounding it. The Saturnian system was a favorite of artists who envisioned what it would be like to travel there: one such example is the popular work by Lucien Rudaux (1874–1947), *Sur les autres mondes* (On other worlds) of 1937. The *Pioneer 11* (1979) and *Voyager 1* (1980) spacecraft brought this fantasy to life and provided more information about the planet and its seven main ring divisions, hundreds of smaller ringlets, and eighteen moons. Astronomers hope to learn even more from the *Cassini* mission, which is expected to rendezvous with Saturn in 2004. The mission is named after the astronomer Jean-Dominique Cassini (1625–1712) who discovered four of Saturn's moons and the eponymous gap in Saturn's rings. *Cassini* is carrying a separate craft, the "Huygens Probe," which will be sent into the methane-rich atmosphere of Saturn's moon, Titan.

Lucien Rudaux. *Sur les autres mondes* (Paris: Librairie Larousse, 1937). Comparative views of the Sun from the surface of Mars and from Earth. (Huntington Library: 43 R82 1937)

Lucien Rudaux. *Sur les autres mondes* (Paris: Librairie Larousse, 1937). Rudaux claimed that clouds on Mars, because of their great height, would shine brightly in the night sky long after the Sun had set. (Huntington Library: FOLIO QB 43 R82 1937)

Lucien Rudaux. *Sur les autres mondes* (Paris: Librairie Larousse, 1937). An artist's conception of Saturn's impressive appearance from one of its nearby moons. The rings of Saturn are viewed edge-on in this scene and appear as the thin streak going from upper left to lower right. (Huntington Library: FOLIO QB 43 R82 1937)

Lucien Rudaux. *Sur les autres mondes* (Paris: Librairie Larousse, 1937). Rudaux's view of a craft entering the atmosphere of Venus. (Huntington Library: FOLIO QB 43 R82 1937)

Lucien Rudaux. *Sur les autres mondes* (Paris: Librairie Larousse, 1937). View of the lunar surface as a landing craft would view it. (Huntington Library: FOLIO QB 43 R82 1937)

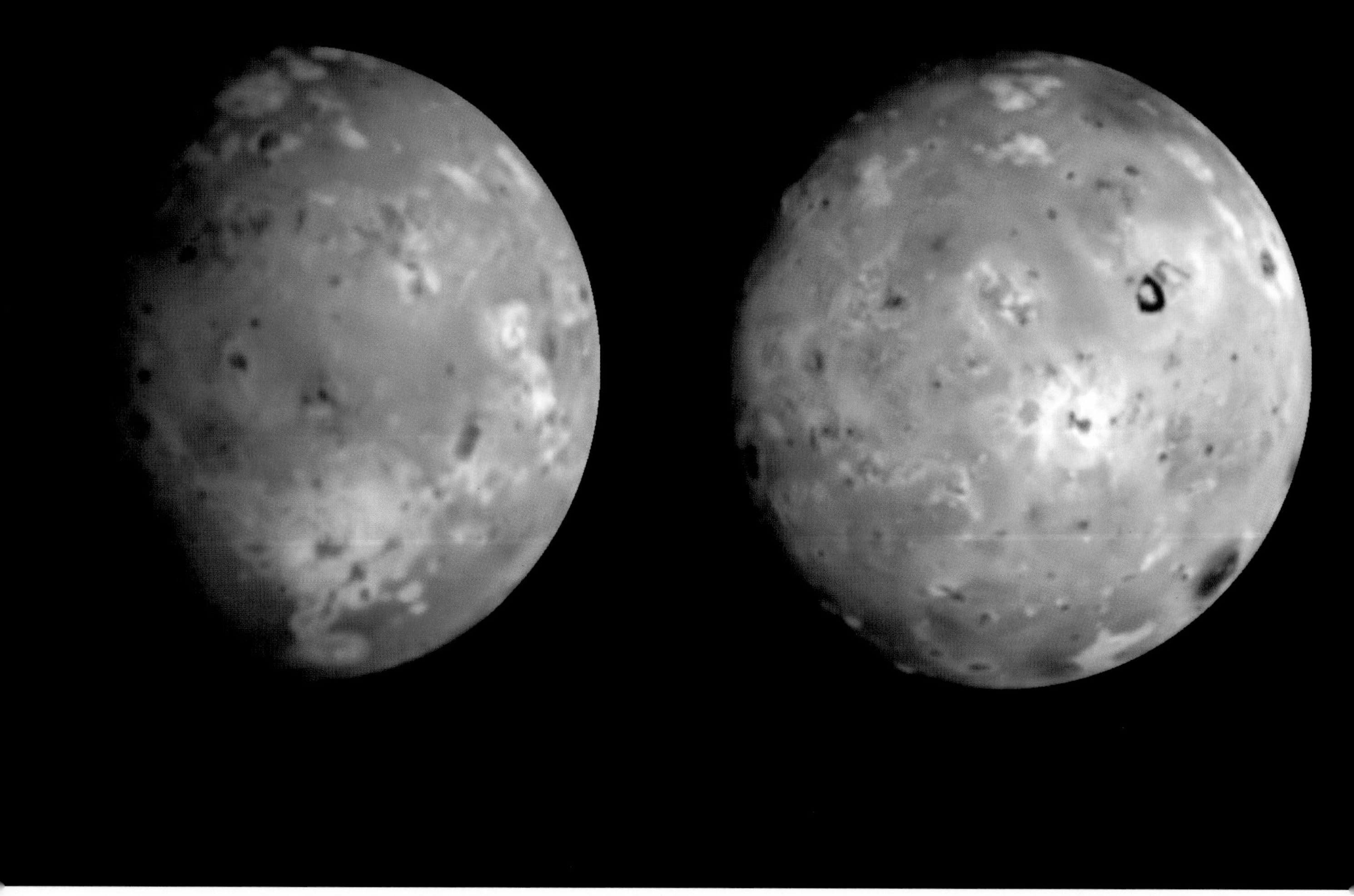

Three full-disk color images of Io, one of Jupiter's moons. (NASA/JPL)

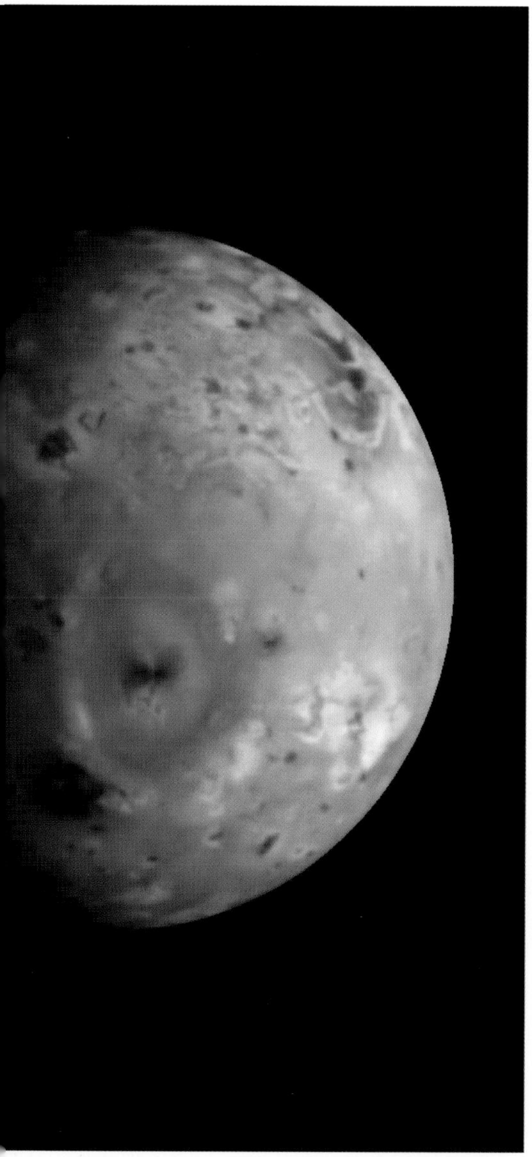

For centuries, these five planets and the Earth, Moon, and Sun made up the solar system. But as telescopes and celestial mechanics improved, new planets were discovered. Uranus was found by accident by William Herschel in 1781, while Neptune was discovered in 1843, in large part because its existence had been predicted by Urbain J. J. Leverrier (1811–1877) and John C. Adams (1819–1892). Both planets are gas giants with ring systems and numerous moons (Uranus has twenty-one, the most of any planet). Uranus and Neptune, which remained mysterious bodies for a long time, revealed their secrets to the *Voyager* 2 spacecraft in 1986 and 1989, respectively. The most recent planet discovered is Pluto, found by Clyde Tombaugh (1906–1997) in 1930. While some astronomers may not consider Pluto to be a "proper" planet, rather closer to an asteroid or comet, it has been considered part of the planetary system too long to be demoted now. Its orbit is quite eccentric, and on some occasions (most recently between 1979 and 1999), it actually gets closer to the Sun than Neptune. Pluto has one moon, Charon (discovered in 1978), which is so close to Pluto in size compared to other moons, that astronomers consider the two to make up a dual-planetary system. But we still know very little about Pluto, with most of our information coming from Earth-based telescopes and the Hubble Space Telescope. It will be a while before a spacecraft is sent to explore it and Charon, although NASA is currently studying one possible mission, called the *Pluto-Kuiper Express*, which would reach Pluto sometime after 2020. *RB*

SIXTEEN | BLAZING SWORDS: *Comets in History*

Halley's Comet as taken 8 March 1986, by W. Liller, Easter Island, part of the International Halley Watch, Large Scale Phenomena Network. (JPL/NASA image)

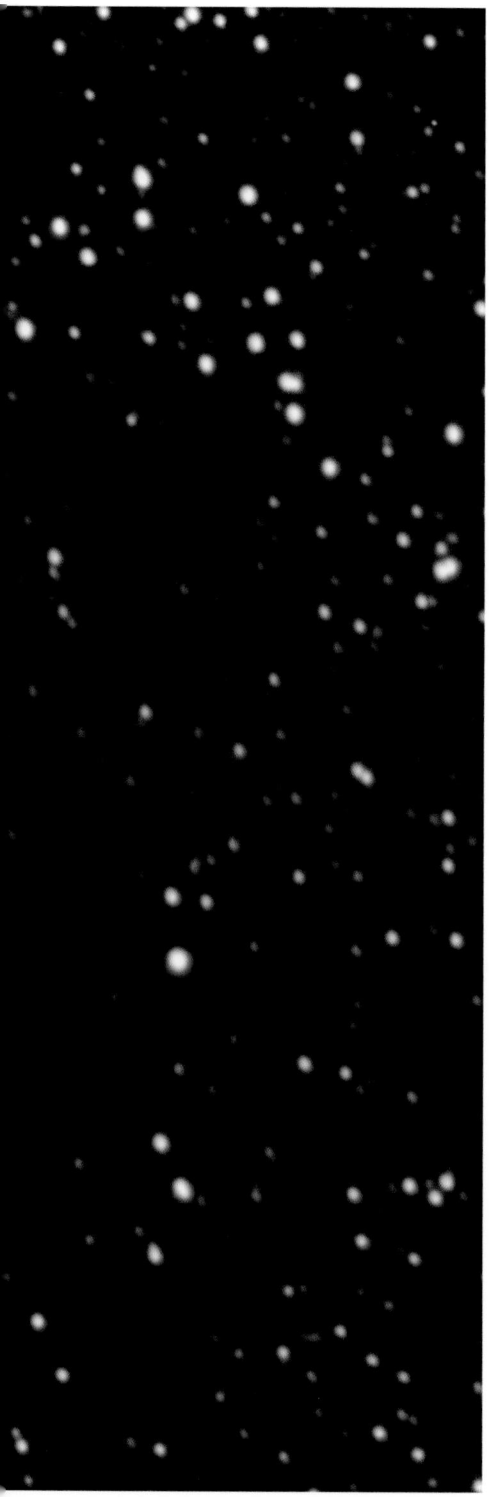

COMETS HAVE BEEN KNOWN AS "BLAZING SWORDS," or often, "bearded stars" for centuries.[10] The word has its origins in the Greek *kometes*, meaning "long-haired." Comets are formed when dust particles embedded in ice form a solid nucleus up to miles across, then approach the Sun. Solar heat vaporizes the ices (consisting of both frozen water and frozen gases) into gas, and the gases expand away from the nucleus to create a gigantic cometary atmosphere. The solar wind forms a dust tail that points away from the Sun. Despite the relatively small size of the body of the comet, the atmosphere and tail of a comet can be spectacularly long.

Comets have been observed in the skies and illustrated by artists since antiquity. The Bayeux Tapestry, an enormous work of embroidery on wool done in the eleventh century not only commemorates the Norman Conquest of England in 1066 but also depicts Halley's Comet, which was most recently visible in March 1986 and is next scheduled to return in 2061. The illustration is the first known of that comet (although the Chinese had noted its existence as early as 240 B.C.E.) As the planet's population has grown and books have become a more reliable and widespread method of transmitting information, we know of more observed comets. Although a total of twenty-seven "great comets" (visible to the naked eye in the night sky) have appeared in the past 450 years, very few are periodic comets—ones with predictable ellipses. The best-known of these periodic comets is Halley's, named after Edmond Halley (1656?–1743), who successfully predicted its return based on the Sun's gravitational pull. There are many known periodic comets besides Halley's. Unlike Halley's, most of these comets lie within the orbit of Jupiter and travel in nearly circular orbits and do not approach the Sun too closely. As a result, they are not very spectacular, and their much smaller tails render most of them invisible to the naked eye. Comets have orbital periods that range from thousands to millions of years.

Isaac Newton played an important role in improving our understanding of the paths comets take on their journeys through space. He developed a technique to compute the parabolic orbit of comets based on viewing a comet on three successive occasions as it moves across the sky. He showed that comets were not erratic interlopers into our solar system but arrived on a schedule and followed a predictable orbit. "I am out of my judgement," he wrote, "if they are not a sort of planets revolving in orbits returning into themselves with a continual motion." He provided a masterful analysis of comets in Book III of his 1687 work *Principia Mathematica*, and helped to undermine the centuries of suspicion and superstition that had accompanied the appearance of comets.

Infrequent and unpredictable visitors, the arrival of comets caused great consternation and excitement. In the days before Halley's work in the late seventeenth century, along with the work of Isaac Newton, comets were almost always considered bad omens, signifying coming disaster. They were terrifying, for people did not have the consolation of science to explain what they meant or why they were there. Aristotle theorized that comets were not celestial phenomena but rather terrestrial, arising when then Sun warmed the Earth, causing evaporation of warm dry matter that rose to the heavens, was ignited by friction, and was then carried around the Earth by the circular motions of the perfectly spherical heavens until they burnt out. This fanciful view went unchallenged for more than two millennium.

Newton's work was followed by a number of other works, most notably a flurry of items produced to coincide with the coming of Halley's Comet in 1757. Benjamin Martin's book *The Theory of Comets, Illustrated, in Four Parts* was published in London the year of the comet's appearance. John Cowley's *Discourse on Comets* appeared the same year. Other popular works continued to appear, such as O.M. Mitchel's 1862 work *The Planetary and Stellar Worlds: A Popular Exposition of the Great Discoveries and Theories of Modern Astronomy.* Mitchel was a very popular writer on astronomy and later went on to serve as an important Union general in the Civil War, continuing to write and soldier simultaneously. He died in battle in the fall of 1862.

Ironically, as knowledge about comets and their nature has accumulated, so too has the growing realization that because they followed erratic elliptical orbits in the solar system, it was possible that they could collide with Earth—as they had done in millennia past. As the Canadian comet discoverer David Levy (of Comet Shoemaker-Levy 9 fame) wryly noted, "Comets are like cats. They have tails, and they do precisely what they want." The Shoemaker-Levy 9 Comet collided with Jupiter's dense atmosphere in 1994, breaking into twenty-one pieces and causing fireballs larger than Earth. This event focused attention on the dangers posed to the Earth by cometary collisions. As of 1995, 878 comets have been cataloged and their orbits at least roughly calculated. Of these, 184 are periodic comets (having orbital periods of less than 200 years); some of the remainder are doubtlessly periodic as well, but their orbits have not been determined with sufficient accuracy. No known periodic comets are on a collision course with Earth. Their tailless cousins asteroids have been pointed to as having possibly been involved in massive collisions with earth some 65 million years ago, with a resultant massive loss of life on the planet.

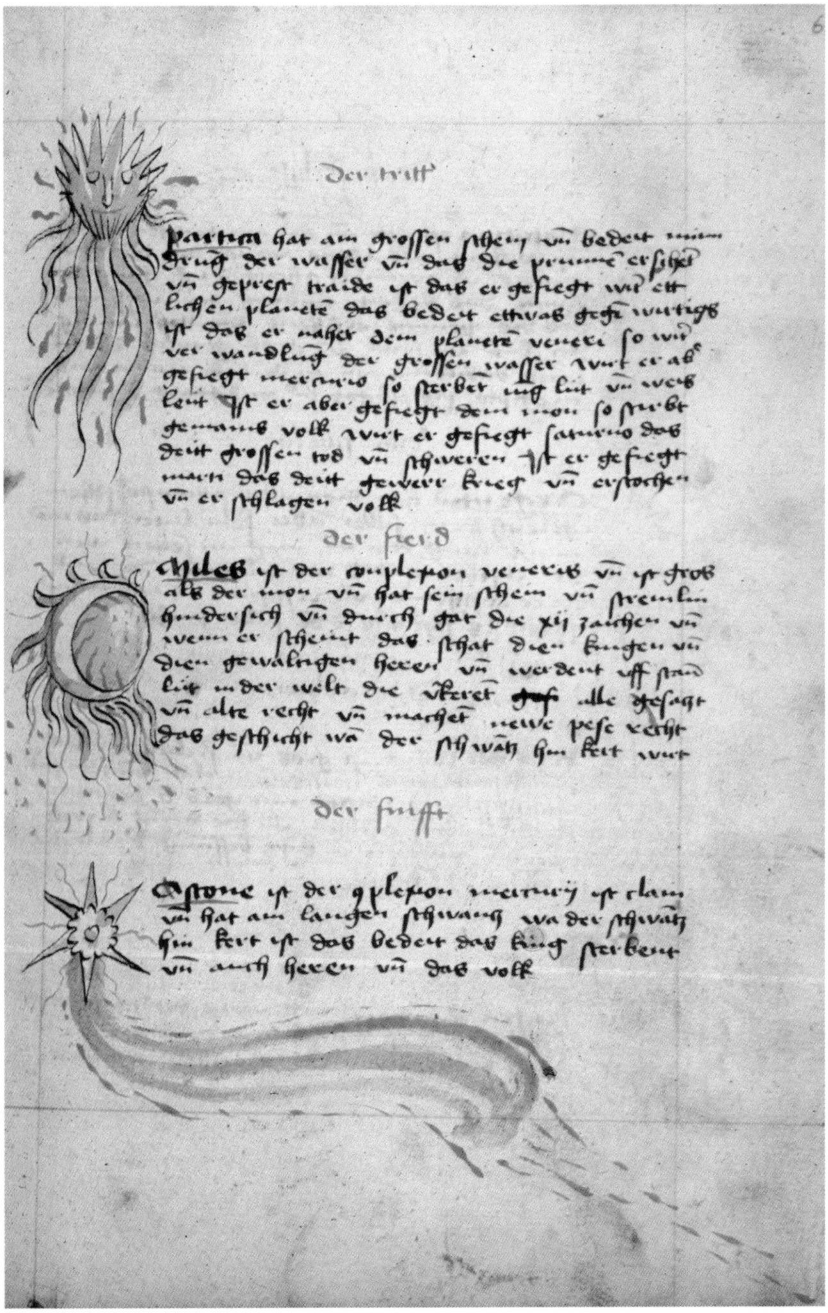

Folio 63 from Ludwig XII/8 Manuscript, ca. 1450–1475, images of comets. From antiquity through the Renaissance, people commented in texts how blazing stars somehow seemed conjoined with such menacing events as plagues. (The J. Paul Getty Museum, Los Angeles).

Despite scientific gains in understanding the basic nature of comets, writers have regularly insisted on the comet as a catastrophic phenomena, one that had historically caused everything from the Great Flood to every major war and conflict experienced on the planet. Ironically, comets are now pointed to in legitimate scientific circles as possibly having played crucial roles in the evolution of the Earth

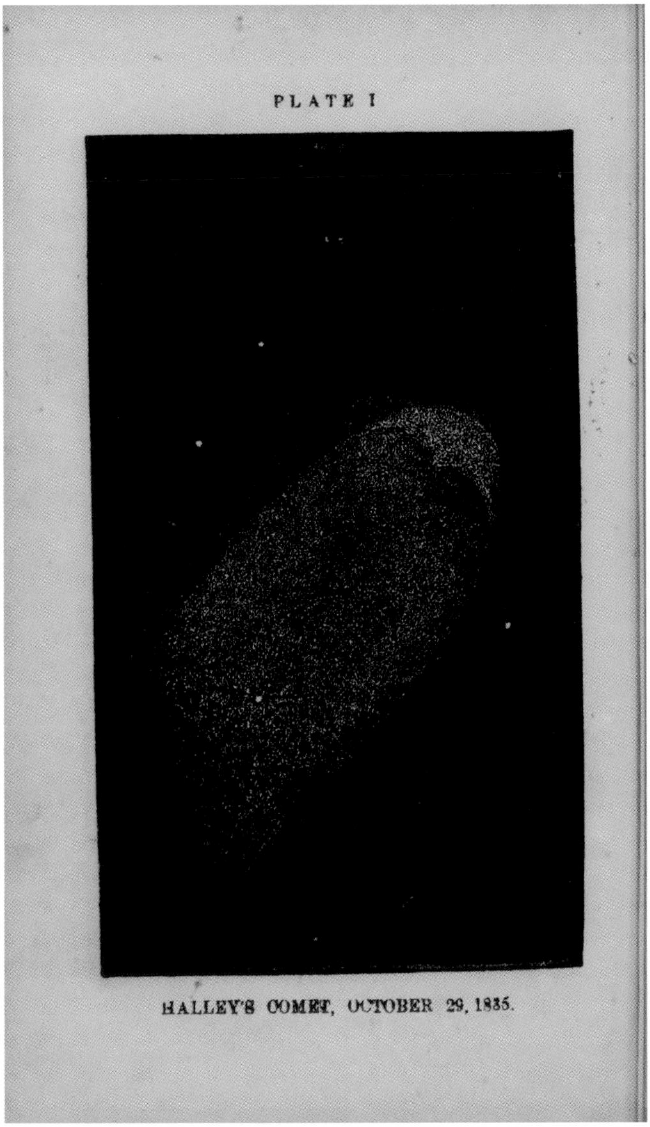

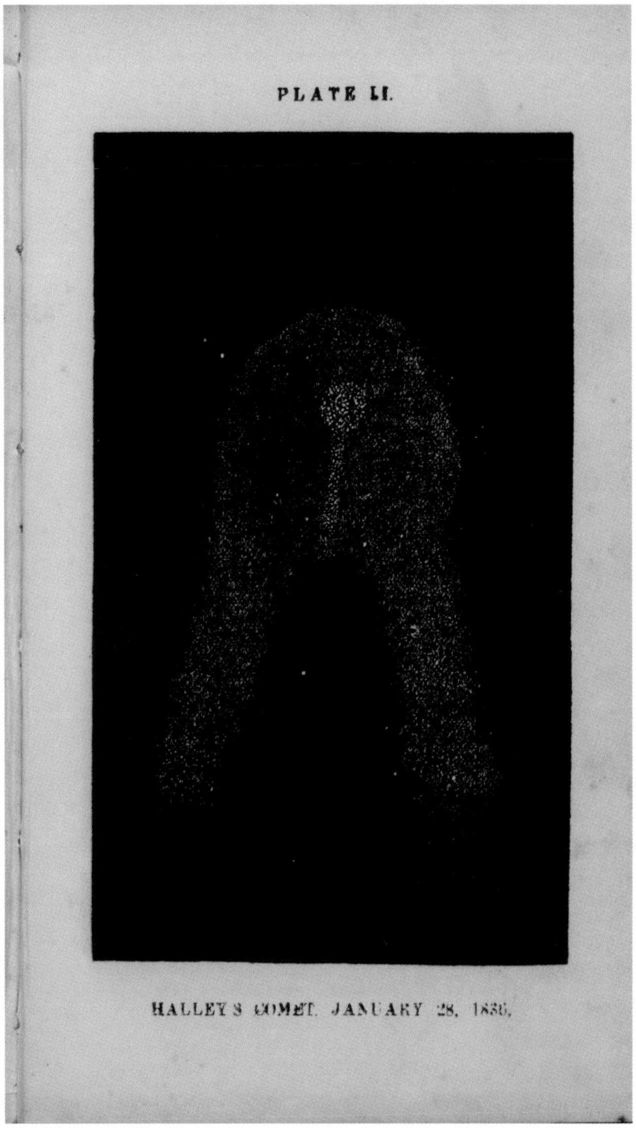

Ormsby M. Mitchel. *The Planetary and Stellar Worlds: A Popular Exposition of the Great Discoveries and Theories of Modern Astronomy* (Glasgow: William Collins, 1862). Halley's Comet has long captured imaginations. Mark Twain, productive during the Civil War when these images were printed, predicted in 1909 that he would die when it returned: "I came in with Halley's Comet in 1835," he remarked. "It is coming again next year, and I expect to go out with it.…The Almighty has said, no doubt: 'Now here are these two unaccountable freaks; they came in together, they must go out together.'" He was right. When Mark Twain died on April 21, 1910, Halley's Comet was once again visible in the sky. (Huntington Library: RB 487000.832)

after all. The *Voyager* spacecraft returned images of deeply cratered landscapes throughout the outer solar system, which resulted in new theories about the significant evolutionary role comets and asteroids may have played in the solar system. Scientists have suggested that comets influenced the course of life on Earth through extinction due to massive collisions of comets with the Earth. Some scientists have also theorized that comets were responsible for depositing the water- and carbon-bearing molecules that led to the formation of life on Earth. They are also important scientifically as samples of the material that formed the primitive solar system. *DL*

SEVENTEEN | THE MILKY WAY AND OTHER GALAXIES

1995, the magnificent spiral galaxy, NGC 4414, was
naged by the Hubble Space Telescope as part of the
ST Key Project on the Extragalactic Distance Scale. An
ternational team of astronomers was led by Dr. Wendy
eedman of the Carnegie Institution of Washington.
Hubble Heritage Team, AURA/ STScI/ NASA)

WE HAVE ONLY KNOWN GALAXIES for a short time. To ancient astronomers, the Milky Way galaxy was just a hazy band in the night sky, milky in appearance (the words galaxy and galactic derive from the Greek and Roman words *gala* and *lac*, for "milk"). Galileo's observations of the Milky Way with a telescope in 1610 showed that this milky stripe across the sky was actually composed of innumerable faint stars. Shortly thereafter, other astronomers found faint nonstellar patches of light scattered about the sky and called these *nebulae*, Latin for "mist" or "cloud." While there was some speculation about the nature of these nebulae, little actual research was performed upon them.

In the late eighteenth century, William Herschel studied these faint nebulae from the perspective of a natural historian, attempting to catalog all the nebulae he could see with his telescopes. Herschel also did star counts of the Milky Way in his groundbreaking study of our galaxy's structure published in 1784 as "Account of some observations tending to investigate the construction of the heavens," in the *Philosophical Transactions of the Royal Society of London*. In this work, Herschel postulated that the solar system was located inside a layer of stars, which caused the Milky Way galaxy to appear the way it does. He continued by declaring his belief that the nebulae are themselves collections of stars like the Milky Way and not just some strange luminous fluid. Herschel had to make some bold and often inaccurate assumptions in his investigations, but they were necessary for him to develop his theories beyond those of his predecessors. Herschel could not, for example, count all the stars nor measure their actual distance, so he assumed that if stars were the same average brightness, a large number of stars in a telescope's field of view meant that the Milky Way galaxy extended farther in that direction than in a field where there were fewer stars. These *gages* (star counts) enabled Herschel to map out a cross-section of a lens-shaped Milky Way with the Sun near its center.

The Anglo-Irish astronomer William Parsons, Third Earl of Rosse (1800–1867), constructed two massive reflecting telescopes in Ireland, one with a 36-inch-diameter mirror and the other with a 72-inch mirror. Lord Rosse and his astronomers studied the nebulae cataloged by Herschel and his son John F. W. Herschel (1792–1871). While they did not detect the nebulae's constituent stars, they did discover that many of them had a spiral structure. Lord Rosse published a number of drawings of these spiral nebulae in an 1850 article, "Observations on the nebulae," which also appeared in the *Philosophical Transactions*. His work firmly established spiral nebulae as a distinct form of nebulae, and the ability of his large telescopes to distinguish individual stars in some of the larger ones helped strengthen the belief that spirals were galaxies (sometimes called "island universes") like our own Milky Way.

In the late 1800s, however, evidence began to accumulate that countered the notion of spiral nebulae being galaxies. The spectra of some nebulae had patterns similar to those of a luminous gas, not clusters of stars. By 1900 astronomers began to doubt the island universe theory and considered the universe and the Milky

Like most other nebulae, the Lagoon Nebula contains clouds of cold molecular hydrogen that are being ionized by a powerful nearby star. In this case the star, O Herschel 36 (seen in lower right), is boiling away the outer layers of the nebula. This "photoevaporation" is causing the blue "haze" around some of the clouds as the heated gas flows away from the cool interior. (AURA/STScI/NASA image)

NGC 4603, a galaxy with majestic spiral arms and intricate dust lanes, is 108 million light-years away. Its distance has been accurately measured by astronomers using one of the fundamental yardsticks of the extragalactic distance scale—pulsating variable stars known as Cepheids. Though intrinsically very bright, Cepheids are faint and difficult to find at such large distances. Thanks to the Hubble Space Telescope's sharp vision, more than thirty-six beckoning Cepheids have been identified in NGC 4603, now the most distant galaxy in which these stars have been located. (AURA/STScI/NASA image)

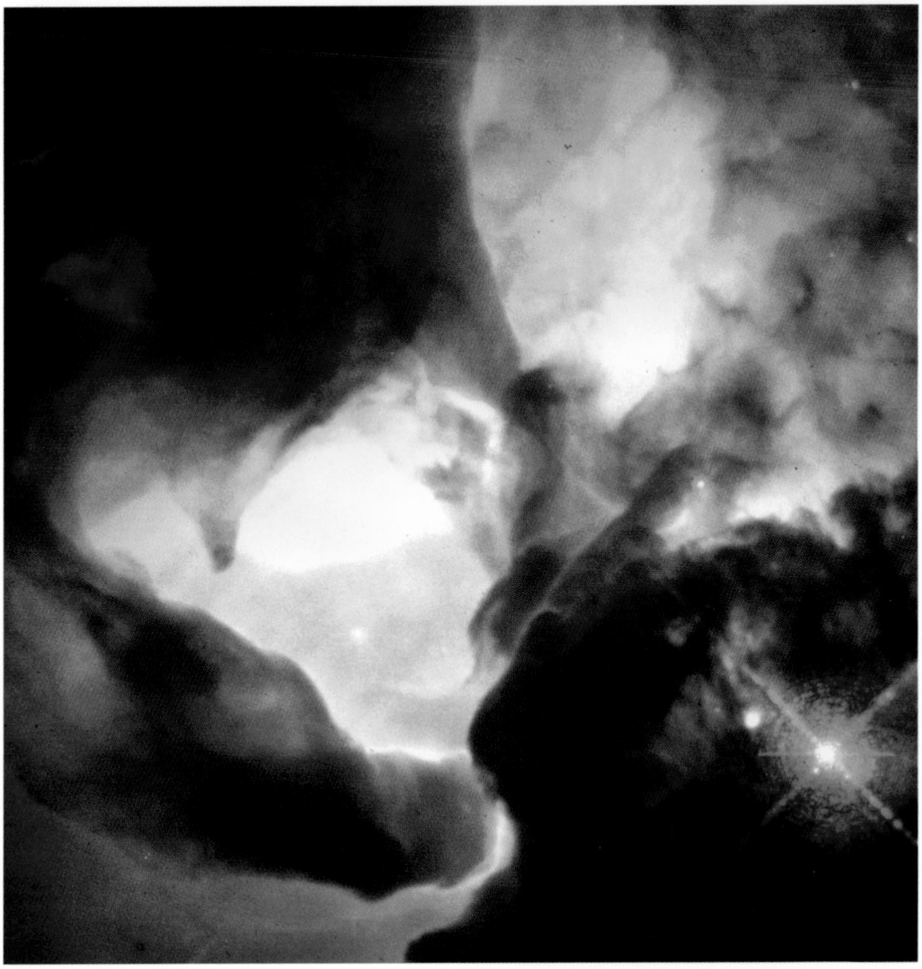

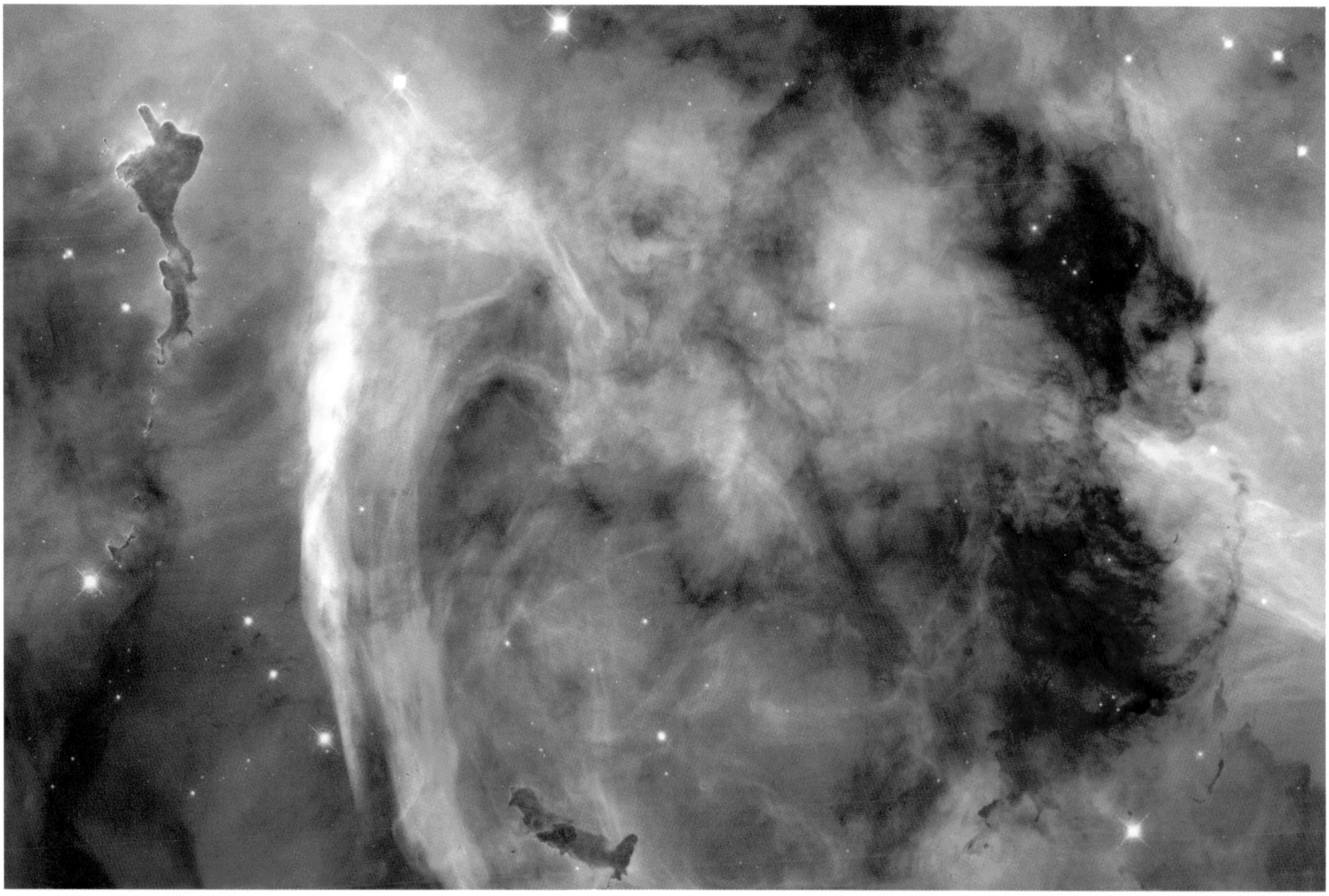

Previously unseen details of a mysterious, complex structure within the Carina Nebula are revealed by this image of the "Keyhole Nebula," seven light-years in diameter, obtained with the Hubble Space Telescope. This montage was assembled from four different views from April 1999, using six different color filters. The Carina Nebula, with an overall diameter of more than two hundred light-years, is one of the outstanding features of the Southern Hemisphere portion of the Milky Way. (The Hubble Heritage Team, AURA/STScI NASA)

Way to be essentially the same. The Milky Way galaxy (or simply, "the galaxy") was considered to be an immense system of stars and there were few who believed that there were similar galaxies beyond its confines, although it was possible that there was an infinite universe beyond the galaxy, forever beyond our observational abilities. By 1910, though, there was some new evidence that indicated that spiral nebulae might truly be far distant star systems like our galaxy. Another major breakthrough in observational cosmology came in 1912 when Henrietta Leavitt (1868–1921) announced that a certain type of star, a Cepheid variable, seemed to vary in brightness over a period of time that depended on its actual luminosity. This period-luminosity relationship, when calibrated, would allow astronomers to calculate the distances to these stars. In 1918 the astronomer Harlow Shapley (1885–1972) calculated that the Milky Way was huge, some three hundred thousand light-years (one light-year being nearly six trillion miles) across. Because of the vast size of the Milky Way, Shapley and others doubted that the spiral nebulae were distant galaxies as they could hardly be so far away as to be outside the galaxy.

Other astronomers were gathering data that suggested that the spiral nebulae were indeed beyond the galaxy, or "extragalactic." In 1923 Edwin Hubble (1889–1953), an astronomer at Mount Wilson Observatory, photographed a nearby spiral nebula, the Great Nebula in Andromeda or M31, and discovered what turned out to be a Cepheid variable. At first he thought it was a nova (a sudden brightening of a previously faint star) and he initially marked the object on the photo-

This stellar swarm is M80 (NGC 6093), one of the densest of the 147 known globular star clusters in the Milky Way Galaxy. Located about 28,000 light-years from Earth, M80 contains hundreds of thousands of stars, all held together by their mutual gravitational attraction. Globular clusters are particularly useful for studying stellar evolution, since all of the stars in the cluster have the same age (about 15 billion years) but cover a range of stellar masses. Every star visible in this image is either more highly evolved than, or in a few rare cases more massive than, our own Sun. (Hubble Heritage Team, AURA/STScI/NASA)

graphic plate with the letter "N" for nova. This photograph, H335H, taken on October 6 with the 100-inch reflecting telescope on Mount Wilson, was of great significance for it allowed Hubble to begin determining the distance to a spiral nebula. Hubble observed this Cepheid for several months and drew out the curve on a chart to see how its light varied. This light curve, as seen in a graphic form drawn in his own hand, enabled Hubble to estimate that M31 was nearly one million light-years away, well beyond the confines of the galaxy and large enough to be a galaxy of its own. Hubble understood the significance of this discovery and returned to his log book of photographic plates and noted the moment. Hubble's work established without a doubt the existence of galaxies beyond the Milky Way and led to the current work on galaxies which shows them to be the basic structural component of the universe. *RB*

In this view, taken in January of 2000, the Hubble Space Telescope has imaged a giant, cosmic magnifying glass, a massive cluster of predominantly spiral and elliptical galaxies some two billion light-years from Earth. The cluster is so massive that its enormous gravitational field deflects light rays passing through it, much as an optical lens bends light to form an image. This phenomenon, called *gravitational lensing,* magnifies, brightens, and distorts images from faraway objects. The cluster's magnifying powers provides a powerful "zoom lens" for viewing distant galaxies—ten to twenty billion light-years away—that could not normally be observed with the largest telescopes. (NASA, A. Fruchter and the ERO Team, STScI, ST-ECF)

Acknowledgments

This catalog, and the Huntington Library exhibition that accompanies it, have been highly collaborative and integrative efforts. The show, *Star Struck: One Thousand Years of the Art and Science of Astronomy,* has allowed us to integrate some of the most beautiful visual imagery from a millennium of representation of the heavens and shows the sometimes-spectacular intersections between art, science, and technology in the pursuit of a greater understanding of the heavens.

While most of the materials in the show have come from the Huntington's own magnificent collections, a number of institutions have loaned objects for exhibit. The J. Paul Getty Museum provided three early astronomy-related manuscripts. The California Institute of Technology loaned a telescope owned by a young George Hale. Telescope manufacturer Celestron provided the use of a beautiful fourteen-inch telescope. The Jet Propulsion Laboratories (JPL) provided expertise and a number of high-resolution modern images. Jurrie J. van der Woude, JPL's image coordinator, provided exemplary sources and aid. NASA loaned us a moon rock for display in the exhibition. Other items in the show come from the holdings of the Carnegie Institution of Washington (CIW), some of which are on permanent deposit at the Huntington Library. John Grula, Librarian at the CIW, as well as Director Gus Oemler and astronomer Alan Sandage, were of great help in obtaining the special items we asked to borrow.

Given the compressed nature of the work (less than ten months from start to finish for both catalog and exhibit), the endeavor would not have taken place without help from many Huntington staff members. Director of Research Roy Richie cajoled, encouraged, and supported our curatorial efforts and contributed to the exhibit's intellectual content. His assistant Carolyn Powell provided additional administrative help to us both. Director of the Library David Zeidberg and his assistant Melanie Pickett handled key administrative and logistical tasks. Mary Robertson, Chief Curator of Manuscripts, provided outstanding support, advice, and succor, as did Peter Blodgett, fresh from a major exhibit of his own here. Olga Tsapina provided relief from many of Dan's regular duties, and the Stacks Supervisors in both the Rare Books and Manuscripts departments—Lisa Libby and Christine Fagan—handled countless requests promptly and cheerfully. Lita Garcia, Kristin Cooper, and others in the Manuscripts Department provided assistance as well. In Conservation, Shelly Smith and Betsy Haude demonstrated patience and expertise. In the Huntington's Image Lab, John Sullivan and his department showed technical derring-do in providing images of the highest quality for numerous uses. Jennifer Sullivan digitally photographed all of the Huntington images that appear in this catalog. The Development Division, including Marilyn Warren, Peggy Spear, Catherine Babcock, Lisa Blackburn, Morgan Kearns, Gina DiMassa, and Randy

Shulman, raised the monies for creating, designing, and mounting the show. Peggy Park Bernal, Director of the Huntington Library Press, was key in the creation of this catalog, which was produced by Perpetua Press, Los Angeles. Editor Tish O'Connor and designer Dana Levy provided skilled and even-handed work. We were also fortunate to have Rob Ball as the lead designer for the exhibit itself. He brought formidable experience and talent, and IQ Magic—Valerie Herzberg and Ragna Jacobsen in particular—provided great creativity in the exhibit's design and production.

Our curatorial efforts with the exhibition and catalog both also benefited from the experience of a trio of experts: historians of science Robert Westman of the University of California, San Diego, and Barbara Becker of the University of California, Irvine, and astrophysicist Michael Werner of the Jet Propulsion Laboratory.

Finally, Dan thanks Pamela Bailey for her love, support, and sense of humor. Her own acute curatorial eye and experience with exhibits helped to inform Dan's views about exhibitions and publications in substantial ways.

Ron expresses his appreciation to the Smithsonian Institution Libraries for their generosity in allowing him time to work on the exhibition in San Marino. He thanks his wife Madeline Copp for her love, support, and patience as he spent a good deal of time away from home at the Huntington and much of his time at home working on the exhibition and talking to himself.

RONALD BRASHEAR, Dibner Library, Smithsonian Institution
DANIEL LEWIS, PH.D., Huntington Library

Index

Page numbers in **boldface** indicate illustrations.

STAR STRUCK
ONE THOUSAND YEARS OF THE ART AND SCIENCE OF ASTRONOMY
was produced by Perpetua Press, Los Angeles
Designer: Dana Levy
Editor: Letitia Burns O'Connor
Indexer: Kathy Talley-Jones
Typeset in Weiss and Aquinas
Printed in Hong Kong
by Toppan Printing Company